T0245115

SOLAR ENERGY:

LET THE SUN SHINE IN

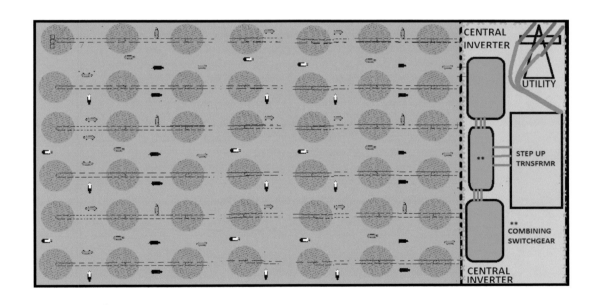

Michael Graham M. Sc.

© 2023 Michael Graham M. Sc.. All rights reserved.

No part of this book may be reproduced, stored in a retrieval system, or transmitted by any means without the written permission of the author.

AuthorHouse™
1663 Liberty Drive
Bloomington, IN 47403
www.authorhouse.com
Phone: 833-262-8899

Because of the dynamic nature of the Internet, any web addresses or links contained in this book may have changed since publication and may no longer be valid. The views expressed in this work are solely those of the author and do not necessarily reflect the views of the publisher, and the publisher hereby disclaims any responsibility for them.

This book is printed on acid-free paper.

ISBN: 979-8-8230-0241-7 (sc)
ISBN: 979-8-8230-0242-4 (e)

Library of Congress Control Number: 2023903758

Print information available on the last page.

Published by AuthorHouse   03/16/2023

authorHOUSE®

CONTENTS

DEDICATION

I dedicate this work to the two Sun's that shone onto my solar collector for 41 years and dislodged my valent electrons to create the energy flow of learning that still guides all my life, Toby and Vera Graham.

MICHAEL ANTHONY GRAHAM

ABSTRACT

This great body through its great magnetic and gravitational pull holds all the planets and their related planetary bodies in their respective orbits within the boundaries of this solar system and the great volume of energy dispensed by the Sun throughout the solar system maintains the perpetual motion, correct interplanetary distances of the planets and life on Earth. This energy generated within the Sun through nuclear fusion (Hanania et al 2020) is not unique to the Sun in our Solar System as there are billions of other such stars throughout the Milky Way Galaxy and the universe (NASA) that have similar energy profiles and planets that orbit them that could similarly help to foster life in many other places if solar energy was all that was required for life.

INTRODUCTION

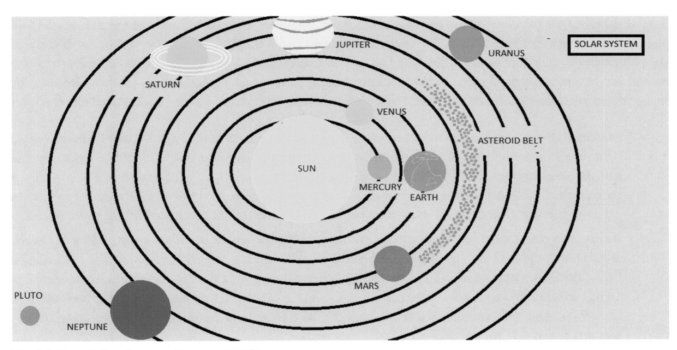

Fig I.1 Our Solar System. Source: NASA

The Earth exists as an integral part of a solar system that is a part of the Milky Way Galaxy which is one of many galaxies in a vast universe that is continually expanding and this vast universe consists of many solar systems and galaxies that are too numerous to count as there are billions of stars filling our universe. This makes the solar system in which the Earth exists relatively minute and physically insignificant except for the fact that this solar system and this galaxy is the only one in which life as humans know it is known to exists, even after many decades or maybe even many centuries of searching across the universe to find other life forms. The significance of this fact is that planet Earth with its varied life forms may be the only one of its kind in the universe and this awesome reality will not be diminished by much even if another planet which is an exact replica of Earth is located somewhere out there with life forms very similar to that on Earth, as this will only make Earth one of two in the universe of billions instead of only one in the same universe.

The Solar System in which the Earth exist consists of eight planets, many moons and an asteroid belt, all of which travel in different elliptical paths around a massive central yellow star which emits life giving energy on which all the planets within the system depend. In this solar system all of the planets rotate around the yellow star in a given path called an orbit, while the whole Solar System rotate around the Milky Way Galaxy. Each planet rotates around the yellow star, called the Sun, in a given orbit, at particular travel times and at set distances from the Sun and the travel time varies from 88 days for the smallest planet to orbit the Sun, to 225- 250 million earth years for the whole Solar System to orbit the Milky Way Galaxy (Earth-Sky 2016). The major elements of the Earth's Solar System are the Sun and eight planets, namely Mercury, Venus, Earth, Mars, Jupiter, Saturn, Uranus, Neptune, the moons that serve these planets and an asteroid belt. Each of the planets have the following characteristics (NASA):

- **Mercury**. This the smallest and the fastest planet with a radius of 1,516 miles, requiring only 88 Earth days to orbit the sun and it has the following characteristics, it is a terrestrial planet that is 36 million miles from the sun, day time temperatures of 430^0 C, night time temperatures of -180^0 C, it has no moon and no rings, has a rough rocky surface that is covered in craters and basins caused by meteoroid and comets crashing into it, it has a crust of basalt and silicates, a relatively large core made up of molten iron and an unbreathable atmosphere.

- **Venus**. This planet has a very thick atmosphere around it that acts to retain heat thereby making Venus the hottest planet and the basic characteristics of Venus are as follows, it is a terrestrial planet that is 67 million miles from the sun, has a radius of 3,760 miles, takes 225 Earth days to orbit the sun, has a surface temperature of 465^0 C, spins backwards relative to the Earth (sun rises in the west and sets in the east), has no moons or rings, has a core of iron, a mantle of molten rock, a crust of thick rock and a surface made up of craters and could not support life due to its high temperatures and acidic clouds.

- **Earth**. For human existence this is the most favorably placed planet relative to the Sun and is the only planet with surface water and the ability to sustain life as it is known on planet Earth and the basic characteristics of earth are as follows, it is a terrestrial planet that is approximately 93 million mile from the sun, it has a radius of approximately 3,958 miles, it takes 365 days to orbit the sun, has surface temperatures that vary between -126 and 136 degrees F, has 1 moon and no rings, has an inner core of iron and nickel, an outer core of iron and nickel fluids, a mantle that is 1800 miles thick consisting of a hot mixture of molten rocks, a crust that varies between 19 and 3 miles at average land levels and the bottom of sea floors, a surface covered with mountains, valleys, plains, lakes, rivers, streams, ponds, volcanoes and a lithosphere crust plus upper mantle made up of moving plates and a protective shield in the atmosphere around it that breaks up meteoroids that enter the atmosphere before they hit the surface.

- **Mars**. One of the most intriguing planets with its proximity to the earth and the possibility that it may have and might be able to support life, as billions of years in the past the planet was both wet and warm unlike its current status. The basic characteristics of Mars are as follows, it is a terrestrial planet that is approximately 142 million miles from the Sun, has radius of 2106 miles, takes 687 Earth days to orbit the sun, has surface temperature that varies between -243 to 68 degrees F, has two moons and no rings, has a core of made up of iron, nickel and sulfur, a mantle of rocks which is between 770-1200 miles thick, a crust 6- 30 miles thick made up of iron, magnesium, aluminum, calcium and potassium and a surface that is rocky, dusty and red in appearance.

- **Jupiter**. The largest of the planets in the solar system with a very interesting visage, specifically it has a permanent large dark red storm, that is larger than the earth, located within it and the basic characteristics of Jupiter are, it is a Gas Giant, that is 477 million miles from the sun, has radius of 43,440 miles, takes 4,333 Earth days to orbit the sun, has a surface temperature of – 234 degrees F, unknown core, no crust, no solid surface, 75 moons and rings and an atmosphere made up of hydrogen, helium, ammonia, ammonium hydrosulfide and ammonium sulfide and does not support life.

- **Saturn**. This is a planet with large rings around it and the rings are the brightest, most massive and complex ring system of any planet and the basic characteristics of Saturn are, it is Gas Giant, that is 886 million miles from the Sun, has a radius of 36,183 miles, that takes 10,756 Earth days to orbit the sun, has a surface temperature of - 284 degrees F, a core that has never been observed, no crust or solid surface, has 53 confirmed moons, 29 others yet to be confirmed and rings, and has an atmosphere made up of ammonia clouds that does not support life.

- **Uranus**. One of the larger planets that is also quite interesting as it appears to tipped on its axis and the basic characteristics of Uranus are, it is a blue green Ice Giant, that is located 1.8 billion mile from the sun, with a radius of 15,759 miles, that takes 30,687 Earth days to orbit the sun, has 27 moons and 13 rings, has a surface temperature of – 356.8 degrees F, has a core of rocks and ice, has no crust, has a swirling liquid surface that turns to ice as it moves toward the core and an atmosphere made up of molecular hydrogen, atomic helium, methane, water and ammonia, that does not support life.

- **Neptune**. This planet is relatively large and has the distinction of being the only planet that was located first by mathematical calculation rather than by physical observations and the basic characteristics of this planet are, it is a blue Ice Giant, that is located 2,782 billion miles from the sun, with a radius of 15,299 miles, that takes 60,190 Earth days to orbit the sun, it has a surface temperature of – 353 degrees F, has 14 moons and 5 rings plus it has a small rocky core with iron, nickel and silicates, an atmosphere made up of molecular hydrogen, atomic helium and methane, that does not support life.

All eight planets, plus their moons and the asteroid belt, receive energy in several forms inclusive of gravitation pull, light and heat, from the giant star at the center of the solar system and this star, called the Sun, is quite massive relative to all of the 8 planets, 201 moons, ring materials and the asteroid belt that it serves, as it contains 99.8% of the total mass of the solar system (NASA). This great body, through its great magnetic and gravitational pull holds all the planets and their related planetary bodies in their respective orbits within the boundaries of this solar system and the great volume of energy dispensed by the Sun throughout the solar system maintains the perpetual motion, correct interplanetary distances of the planets and life on Earth. The energy that is utilized by the Sun is generated within the Sun through nuclear fusion (Hanania et al 2020), however, this is not unique to the Sun in our Solar System as there are billions of other such stars throughout the Milky Way Galaxy and the universe (NASA) that have similar energy profiles and the planets that orbit them could similarly help to foster life in many other places if solar energy was all that was required for life. The role of the Sun within the Solar System could be described as that of a great ring master that must continually rotate eight massive bodies in different planes, while maintaining perfect balance in perpetuity and also providing the required energy to each of these eight bodies need to sustain themselves. This juggling act must be maintained ad infinitum as it is quite probable that any major disturbance to any of these bodies could lead to the utter destruction of all the others including the ringmaster himself.

The energy provided by the Sun to the Earth is much more than the Earth could have ever used and this has always been the case from the early life of the Earth when there was no human, animal or plant life forms, up until the present time when the human population has grown to just over eight billion and is demanding even more energy everyday with its continuously growing numbers. This demand for more energy by humans was met in the past by the consumption of the many great forests which had existed from 400-500 million years ago in time, however, most of these forests were decimated and never regrew leaving barren wasted lands in their wake. The decimation of these forests forced humans to find and develop new energy sources and these new energy sources that were developed during a period of 150 -1000 years have all been proven to be a source of much harm to humans and the environment and they are forcing humans to once again seek and utilize new sources of energy, sources that will have fewer negative impacts on the planet and humans, than the current sources. There are currently several other sources that can be utilized, however, there is one source that is greatly underutilized even though it is more available in greater abundance everywhere than all of the other sources, it is free to use and has no negative impact upon the environment or human, this source is the energy from the Sun. This is a source of energy that has always been essential to life, but was never fully understood or fully utilized by humans, because until about one hundred and fifty years ago humans did not have the technology to fully optimized the energy that emanates from the Sun.

CHAPTER 1

Solar Radiation and Planet Earth

The Earth travels in an "elliptical" orbit around the Sun each Earth year, in approximately 365 Earth days, (Williams 2014) and in this orbit the earth is at one point closer to the Sun than throughout the rest of its orbit and at another point a little further from the sun. The point in the orbit where the earth is closest to the sun is usually referred to as perihelion, which is usually at a distance of approximately 89,729,825 miles (147,098,074 kilometers) from the sun and the Earth is usually in this position on January 3rd each year. The point in the orbit where the earth is furthest from the Sun, called the aphelion, which is usually at a distance of approximately 92,779,170 miles (152,097,000 kilometers) from the Sun and the earth is usually in this position on July 4th each year. When the Sun is in either of these positions, perihelion and aphelion, it will have different impacts upon the amount of solar energy that reaches the surface of the Earth and the difference in solar impact, impacts almost all life on the planet, as the shorter distance could mean warmer weather conditions and the longer distance usually mean colder weather conditions, however, this is not necessarily the case as January 3rd is usually very cold in the northern hemisphere and very hot in the southern hemisphere and July 4th is usually very hot in the northern hemisphere and very cold in the southern hemisphere.

The Sun's impact upon the Earth is a function of several factors inclusive of the tilt of the Earth's axis, 23.5^0 off the vertical relative to the Earth's plane of orbit, the direction in which the axis of the earth is pointed during its travel around the orbit, the distance from the Sun, the points at which the solar radiation most directly strikes the Earth, the equator, and the periodical changes in the Earth's axial tilt, 22.1 -$24.5,^0$ which occurs gradually over time every 26,000 years (Hocken 2019). The axial tilt of the Earth and the angle at which the Earth moves around the sun also determines the climatic conditions at different points on the surface of the Earth in an annual cycle, these climatic conditions include what are known as the four seasonal conditions that impacts the temperate regions of the planet. The conditions in the other regions of the Earth, the North Pole, The South Pole and the Tropical Region are also greatly impacted, with the Tropical region being continuously hot and the two polar regions, north and south, being continuously cold. A temperate region occurs on both sides of the tropical region and the four seasonal conditions also impact the temperate regions in the north and south of the planet at different times during the year, during the Earth's orbit around the Sun. In the northern temperate region, the climatic conditions are cold from September to March and warm from April to August, while in the southern temperate region the climatic conditions are cold from April to August and warm from September until March each year. The different regions of the Earth also receive different amount of solar energy as the Earth orbits the Sun, depending on their latitudinal distance from the equator, with the greatest amount of solar energy impacting the equatorial regions of the Earth and the furthest distances, the North and South Poles receiving the least.

All of the planets are located at different distances from the Sun and the Earth along with the other planets receive different amounts of solar radiation based on their distance from the sun, their size, the planet's atmosphere, the nature of their surfaces and according to the American Chemical Society (ACS) the energy supplied to the four terrestrial planets are as shown in the table below:

Planet	Mercury	Venus	Earth	Mars
D_p km	58,000,000	108,000,000	150,000,000	230,000,000
S_p, W.m^{-2}	9,180	2,650	1,370	580
S_{ave} w.m^{-2}	2,290	662	342	145

Table 1.1 Solar Energy Reaching the 4 Terrestrial Planets. Source: American Chemical Society

The meaning of the abbreviations as indicated in the table above are as follows:

- D_p – The distance of the planet from the sun in kilometers.
- s_p – The energy flux at any place on the surface of the planet
- s_{ave} – The average energy flux over the area of the planet
- $s_{ave} = s_p/4$
- The Earth's energy flux, Sp, has a value ranging from 1370 – 1361 W/m^2, and is usually referred to as the solar constant for earth.

Solar Constant. The Earth like the other planets receive an average amount of solar energy per unit meter square of the Earth surface, this average amount of energy is referred to as the Solar Constant and the definition of the this constant is as follows- The Solar Constant is the radiant energy per cross sectional unit area that the Sun provides for the Earth's systems and can also be described as "the total energy output of the Sun divided by the area (A_{SE}) of the 'Big Sphere' as shown in Fig 1.3 below and as described by the equation - S_o = (3.87 x 10^{26} W)/ A_{SE} = approximately 1368 W/m^2

A_{se} = 4.pi. (R$_{se}$)^2

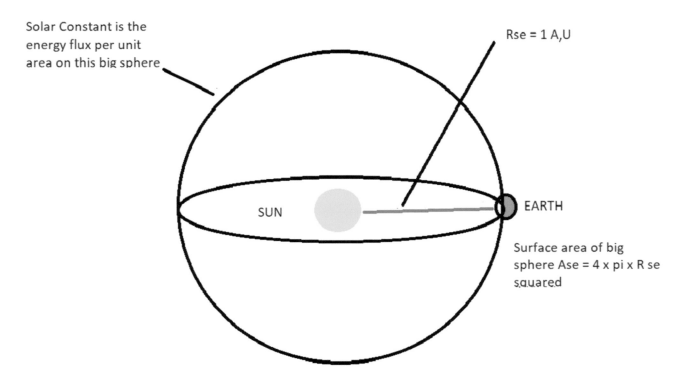

Solar Constant is the energy flux per unit area on this big sphere

Rse = 1 A,U

SUN

EARTH

Rse = 1 A,U

Surface area of big sphere Ase = 4 x pi x R se squared

Fig. 1.1 Solar Constant Big Sphere Diagram. Source: Publicasu.edu

1 AU = 1 astronomical unit, which is equal to 149.6 million kilometers, the mean distance from the center of the Earth to the center of the Sun.

Incident Solar Radiation

While the Earth is supplied with a constant source of energy not all of this incident energy is absorbed or otherwise utilized by the Earth as a fairly large portion of it is reflected,6% from atmosphere, 20% from clouds and 4 %

from the surface of the Earth, a total of 30% of the incident energy is reflected back into space without positively impacting the Earth, leaving a balance of 70% of which 16% is absorbed by the atmosphere, 3% is absorbed by the clouds and 51 % is absorbed by lands and water. A significant portion of this energy that is absorbed by the land, water, cloud and atmosphere is also lost due to reradiation back into space. (Amin et al. 2020). The 51% that is absorbed by the land and surface waters is utilized by the Earth in many ways inclusive of the production of essential plant life, providing the driving force for the hydrologic cycle and the support of animal and human life, however, this 51 % is not always utilized in the most efficient manner and it is possible that with the right technologies much more of the Sun's energy that reaches the surface of the earth could be better utilized to create new motive forces to better serve humans and the environment by replacing the existing energy sources which are currently used to drive industry, transportation, lighting, heating and cooling for industrial, commercial, residential, institutional, health and educational facilities. These other sources of energy are also directly dependent upon the energy from the Sun for their creation over a very long period of time, unfortunately however, these sources were created using the element called carbon, which if wrongly utilized can create many negative impacts on the conditions of the planet.

Historical Use of Solar Energy

Humans have always needed and utilized many things to create the heat and light energy that they needed for their everyday existence and they have also always managed to find new sources of energy when the old source that are easily available have been depleted. However, the depletion of the easily available sources led to humans seeking other sources beneath the soil where they found more consumable material like aged wood and coal, which they developed ways for using and used until demands for energy grew greater and another source of energy was located, this time oil which could utilized much easier than coal. Both coal and oil proved to be more than adequate suppliers of the energy required for all human activities but unfortunately, their use came with a significantly high price, a price that many considered to be far too great as their continued use appeared to be placing significant limitations on human existence. The fear that this realization created caused the human population to once again start the process of finding a newer, cleaner, safer source of energy that could eliminate the what they came to see as an impending crisis and the drive to utilize a cleaner, safer energy led to the vision to utilize a greater portion of the Sun's energy. This current drive to utilize more of the Sun's energy, solar energy, requires the conversion of the Sun's radiation to create electrical energy for use in modern society inclusive of in homes, businesses and industries, however, this desire to utilize more of the Sun's energy directly is not new as in the recent and more distant past natural solar energy had been extensively utilized for almost every human endeavor from the beginning of human existence through to modern times to help meet all basic human physiological needs inclusive of eating, drinking, heating, cooking, sleeping and washing. Humans have utilized direct solar energy in many productive endeavors inclusive of the production of things to eat, agriculture, water to drink, hydrology, the long term preservation of reaped agricultural crops and animal products, the construction of shelters to protect themselves from the elements, residential abode, and the construction of places for communal gathering, religious facilities, the construction of household and other vessels and daylight heating, all of which have been fundamental to the growth and wellbeing of early human lives and civilizations.

Agriculture. The early development of agriculture created the base for the development and growth of human civilizations as the creation of a regular, consistent source of nutrition that the domesticated crops provided required human stability for the processes involved in clearing and preparing lands, planting, reaping, processing and storing of the excess crops for long-term use. This need to be stable caused by realities of agriculture, caused the human population to plants roots in one location and as they did so they created permanent place of residency, places for storing agricultural goods and places of business which were to later become the center of many villages that were started which consequently grew to become major towns, cities and centers of learning. This essential system of agriculture was and is based on the fundamental principle of photosynthesis whereby all plants (trees, shrubs, grass, fungi and vines) utilize solar energy to grow and to produce fruits, seeds, tuber, grains, vegetables and herbs, all of which became essential elements in the diets of early humans starting in their days as Hunter-Gatherers, as humans found these many products to be tasty, safe, healthy and nutritious to consume. In ancient times the Hunter-Gathers sought and located plant products and consumed those that they had developed a taste for, however, with developments in knowledge humans learnt to domesticate the plants that they usually found by chance and propagated them by planting seeds or portions of an old plant and then allowed the combination of sunlight (solar energy), carbon dioxide, water and soil nutrients to do their work. These developments led to the selection of a wider variety of plants for consumption, the planting of larger areas and the need for the organization of labor to prepare the land, plant the seeds, reap, process and store the resulting crops.

These early humans also learnt over time that they needed to process the grain crops to separate the grains from the rest of the plant on which they had grown and they also learnt that the processing could be made easier if the plant matter was dry and that the required drying could be done by the heat Sun. This need to utilize sunlight, solar energy, after the crops had been reaped extended beyond the need to separate the grains from the plant matter as the early humans also learned that the heat from the sun could also to help them to process the grains to make them suitable for storing as these crops, especially the grains usually contained a very high percentage of water, water that facilitated the early spoilage of the crops. The early spoilage of the grains usually meant that the people would be forced to eat grains that had started to deteriorate or had gone too bad to be eaten and this would mean starvation and death until a new crop could be reaped. The discovery that drying the grains could make a difference in how long the grains could last also mean that fewer people would become sick or starve before the next crop was ready. The need for drying utilizing solar energy as the source to dry the crops grew with the growing population and consequently the need to grow and store more crops. The need to grow and store more crops was even more important in the regions of the world that had short growing seasons and long winters in which no crops could be grown.

Food Preservation. The seasonal changes in the temperate regions of the world usually means that certain food crops can only be grown for short periods during the year and to ensure that there will be sufficient food to last the whole year ways had to be found to prevent the crops reaped from spoiling and food preservation and storage became very critical for the next stage in human growth and development. This need for food preservation caused humans to develop two basic methods of preventing food from spoiling and both of these methods were totally dependent upon solar energy. The first method developed, was utilizing the direct heat from the Sun to dry out the natural water contained in all crops when they are just reaped and this drying out allowed for the reaped crops to last much longer. These grain crops were dried by spreading them out thinly in large open, hard areas or areas that contained very little water and leaving them exposed to direct sunlight to allow the heat of the sun to penetrate and drive off the natural water contained within the crops. The removal of the natural water contained in the grains or other produce prevented the growth of the water dependent bacteria and fungi that are responsible for the deterioration and spoilage of crops, thereby extending the life of the crops. The same drying process was also utilized in the preservation of animal products as it also prevented the growth of water dependent bacteria and fungi that leads to spoilage in flesh.

The other method of food preservation that was developed was also dependent upon the use of solar energy as this second method utilized common edible salt for preserving meats, fish and poultry as the addition of this salt to these products ensured that the bacteria and fungi that usually led to the deterioration of the flesh products were destroyed and thereby retarding the deterioration process and consequently extending the useful life of the product. The salt used in this process is usually reaped and dried using the heat of the Sun's radiation to drive off excess water and this drying process was also utilized to form the salt produced into blocks, which was critical for long distance transportation.

Construction. Early humans either lived in natural shelters such as caves or created rudimentary shelters from the natural material around them and some of the rudimentary shelters that these early humans created were constructed from a combination of wet soil, mud, and plant material which when properly combined provided adequate shelter against the natural elements of heat, cold, wind and rain. Generally, these early residences constructed using mud were constructed using wood or reed to form the frame of the building and these materials could be found in the following elements, siding, roofing, door and any other support for the building, while mud combined with grass would be used to cover and seal all the sides and the roof would be covered by grass, or the waste from crops to prevent the entrance of any precipitation, direct sun rays and wind. The mud is normally applied in a wet state and is dried out by the heat of the sun, solar energy, after which it becomes hard, water resistant and more durable against the natural events, human and animal activities. Later growth in construction technology saw the mud converted to blocks and bricks that was used almost exclusively to construct residences, government buildings, temples and other less important buildings and this development in construction technology saw the mud placed in molds to give the mud shape, which is usually square or rectangular, after which these shaped mud forms would be placed in the sun to be dried out and after adequate drying these sun-dried blocks were used to construct residences, large or small, temples, palaces and government facilities. The direct heat of the sun was however, insufficient as it could not chemically transform the clay in the mud to make the blocks hard, durable and long-lasting and the sundried bricks would have to be replaced fairly frequently in the life of a building. Later developments in construction technology saw the elimination of this earlier failure as the block makers realized that to chemically transform the clay greater heat was required in curing of the blocks if the block were to become harder, stronger and more durable and a decision was made to generate the required heat and temperature by burning wood in the presence of the clay blocks to a sufficiently high temperature that could change the nature of the clay so that it became hard and strong. This combination of clay, mud and high heat energy was also responsible for the creation and development of household utensils that became very important in the everyday existence of humans inclusive of cooking pots, storage jars, plates, cups, decorative items, religious items and many other basic household items.

The use of solar energy in early history was not limited to the examples given above and according to Richardson (2018) solar energy was used in many other ways inclusive of the following:

- Humans used solar energy along with magnifying glasses to light fires as early as the 7th century BC.
- The Greeks and the Romans utilized mirrors and solar energy to light torches for religious ceremonies in the 3rd century BC and the mirrors were called "burning mirrors".
- The Chinese also recorded using "burning mirrors" for religious ceremonies in AD 20.
- The Romans and other early civilizations built "sunrooms" on the south facing sides of their buildings to utilize solar energy for home heating.
- The Early Pueblo People of the US built south facing buildings on high cliffs to harness solar heat during the cold months.
- During the late 18th and 19th centuries researchers and scientist utilized sunlight to power ovens and steam boats for long voyages.

The use of solar energy progressed significantly in many different stages through time until the early twentieth century when new technology was developed to utilize the energy of the Sun in a much different way, not directly as in earlier times but indirectly as the energy of the Sun would now be converted to another form of energy that was better suited for the industrial age. This new technology greatly enhanced the use of solar energy as it had the ability to convert sunlight into electrical energy through a device called a "solar cell", the development of which came about as follows:

- French scientist, Edmund Becquerel discovered that light could increase electricity generation when two metal electrodes were placed into a conducting solution and this ability was described as the "photovoltaic effect".

- Scientist Willoughby Smith discovered in 1873 that selenium had photoconductive potential.

- In 1876 William Grylls Adams, and Richard Evans Days discovered that when selenium is exposed to sunlight it generates electricity.

- Charles Fritts created the first solar cell made from selenium wafers in 1883.

- Daryl Chapin, Calvin Fuller and Gerald Pearson created the first silicon photovoltaic cell in Bell Laboratories in 1954 and this new technology had the capacity to power an electric device for several hours each day.

- Solar energy was first used to power satellites in 1958

- The first solar energy system that was used to power a house was created in 1973

- The use of solar energy in a US government facility occurred in 1979 when solar panels were installed at the US presidential residence, the White House.

- The growth in the improvement of solar cells came about in many steps starting from 1873 and the greater improvements were made when the ability of the material used to create the later solar cells, silicon, was modified to improve the ability of the material to better convert incident solar radiation to electrical energy, a parameter called the conversion efficiency. This conversion efficiency saw significant improvements during the following years, in the period 1957- 1960 the conversion efficiency improved from 8-14%, in 1985 the conversion efficiency rose to 20% in work done by University of South Wales, in 1999 the conversion efficiency rose to 33.3% in work done by NREL and 2016 the conversion efficiency rose to 34.5% in work done by University of South Wales.

- During these periods of growth there also occurred a great improvement in the confidence levels relative to the capacity of solar cells, with confidence so high that the first solar powered airplane was created by Paul Macready and it made a debut flight in 1981.

- Confidence in solar cells continued to grow with the flight of another solar powered plane that set an altitude record of 80,000 ft in 1998.

- This confidence in the capacity of solar energy reached its apogee when in 2016 a solar power plane created by Bertrand Picard completed a first round the world trip.

- While it was an impressive technology solar energy would not be of much value if it was not economically feasible to build, install and operate and initial high cost to create install and operate was significant enough to determine the viability of the technology, fortunately however, the viability of solar energy greatly increased with the cost to produce solar energy falling from $300/watt in 1956 to $0.05/watt in the present day and this development makes solar energy competitive with energy produced by fossil fuels.

While the viability of solar energy systems has been established, research and development in solar energy material and systems still continues as there is a great need to further reduce the cost per kilowatt and to find new material that are more environmentally friendlier than silicon, which is currently the material that produces the most efficient solar cells, but it is not environmentally friendly and it is relatively expensive to produce, from mining to purification to the high level, 99.9999 % pure, required to make the most efficient solar cells. Work therefore continues to find and or to create new materials and technologies to replace silicon solar cell technology.

The Other Solar Energies

In most discussions about solar energy the focus is usually on the direct use of the incident solar radiation from the sun in the form of heat or photons of light which must be collected by solar thermal systems or solar collectors which collect and convert incident solar radiation to create electricity. This, however, is not the only form that solar energy takes or is available in on the planet, as all other sources and forms of energy on the planet Earth are sources of energy created by the solar radiation from the sun, as the sun is responsible for the growing plants and animals which produce the fossil fuels, creates the wind that can be used to energize wind turbines, create wave movements that can be used to drive water turbines and creates the hydrological cycle that provide the large volumes of water required for use in hydropower plants to provide the motive force that will energize the hydro-turbines responsible for rotating the electrical generator to create the electrical power.

Hydrology

The most important work performed by the energy from the sun has always been the creation, development and operation of the Earth's hydrological system that is completely driven by energy from the Sun. The heat from the sun provides the energy required to heat and vaporize water from all points on the globe inclusive of all surface water bodies such as oceans, lakes, rivers, streams and ponds, from the soil, all plants (evapotranspiration) and all animals inclusive of humans (perspiration) and to transport this vaporized water from the different locations and deposits this vaporized water into the atmosphere, where it will be held until the atmosphere becomes heavily saturated with this water vapor at which point precipitation of this moisture will take place in the form of rain, snow, hail or ice. This hydrological cycle continues ad infinitum to provide water for use on the surface of the Earth, on land or in water bodies, inclusive of the oceans, lakes, rivers, ponds and in underground aquifers.

Wind

By definition wind is the movement of air and this movement is cause by the different rates at which the Sun heats up different surfaces on the Earth and the fact that nature abhors a vacuum. The heat of the sun usually impacts land and water at the same time but because of the different heat capacities of the land and the sea, the land will heat up much faster than the sea during daylight hours and this means that the air over the land will expand much faster than the air over the sea. Due to convection currents the heated air over the land will expand and rise leaving a vacuum behind which is normally filled by the colder air moving in to fill the vacated space from over the sea. This same result will be repeated with wind at the planetary level as the air over the equator is normally heated up much more than the air over the poles and it will expand much faster leaving much room for the colder air of the poles to move in the fill the vacuum left by the air over the equator. This movement of the air caused by the heat of the Sun is the source of energy for all wind energy systems and it is completely renewable as long as there is a sun providing it energy to the Earth.

Waves

Waves are described as the oscillatory movement of a water surface and are generally caused by a transfer of energy from the source of the waves which can be wind, seismic action or the gravitational pulls of the Sun and moon. In the case of ocean wave there are said to three such major waves types, Wind Waves, Seismic Sea Waves and Tidal wave all of which are the results of different natural forces and are described as follows:

- Wind Waves. The Wind Waves are caused by the action of wind upon the surface of the water body and are usually the most frequent type of the waves.

- Seismic Waves. Seismic Waves or tsunamis are usually caused by the sudden displacement of the ocean floor which is usually the action of earthquakes, landslides and volcanos beneath the ocean or an impact caused by a meteoric strike and these seismic waves tend to have a more devastating effect on land masses and populations within the ranges of these waves due to their long wavelength and the momentum created by the large volume of water that they usually carry with them. These waves will not stop at shorelines but will usually flow very deep inland.

- Tidal Waves. The Tidal Waves are a natural function of the gravitational pull of both the Sun and the moon and centrifugal forces in the solar system, tidal waves usually occur during specific time periods.

With the exception of the seismic waves, the Sun is directly responsible for the other two types of waves, winds and tidal waves, Wind Waves because the Sun is directly responsible for wind and Tidal Waves which are caused by the gravitational pull of the Sun and Moon.

CHAPTER 2

Solar Thermal Energy Systems

Solar energy was utilized in ancient and premodern time in many ways but mainly as a light source for photosynthesis, general life and also as thermal energy and in the modern society it is utilized in these ways but also in many more new ways that have been developed within the last 150 years inclusive of, thermal energy for residential and commercial cooling and heating systems, the drying of agricultural products, in the generation of electrical energy utilizing photovoltaic cells and accessories, to generate electrical energy utilizing concentrated solar power (CSP) systems that are utilized to generate high temperature and pressure steam that may be used in prime movers, to generate of electrical energy utilizing Sterling Engines or similar engines that directly generate electrical energy without the intermediate step of producing steam and it is also utilized in industrial processes to facilitate desired chemical and or physical changes and or transformations.

Solar Thermal Energy Systems

Solar thermal energy systems are designed to utilize the heat available in solar radiation (sunlight) to heat water in domestic hot-water systems, solar thermal electrical power generation plants, to provide heat for industrial processing and to provide heat for cooling and drying systems.

Solar Domestic Hot-water Systems. The basic components that make up a solar domestic hot-water system includes the solar radiation collector, a storage system, a backup electrical heater, a pump or syphon system, temperature sensors and electrical controls, and these domestic hot-water systems are usually designed to operate as either a passive system or an active system. In the passive systems water is circulated through the collector from which it travels to the storage system and later to the point of use by the convection current that are established in a body of hot and cold water, while in an active system the water is circulated at all times by a system of pumps to ensure that hot water will be provided at the outlets at all times. The passive systems, which also include the thermosyphon systems are usually adequate for single family residential use, however, much large facilities such as residential dormitories and hotels usually require the use of pumped systems that will continuously produce hot water at the outlet at all times. The thermosyphon system also required that the storage must be installed above the collector to ensure the desired and unaided flow of water through the system. All of these systems are as shown the Fig 2.1 – 2.3 below.

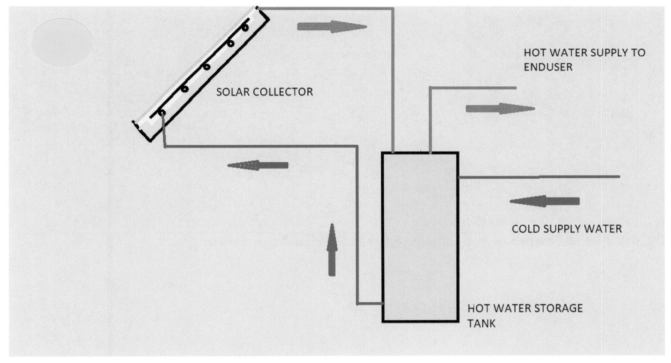

Fig. 2.1 Typical Passive Domestic Water Heating System.

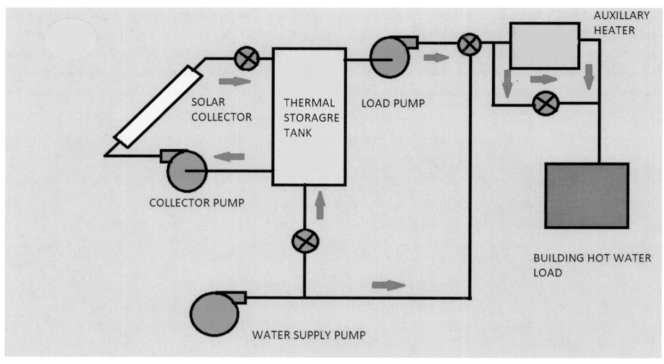

Fig. 2.2 Typical Pumped or Recirculating Solar Hot Water Heater System.

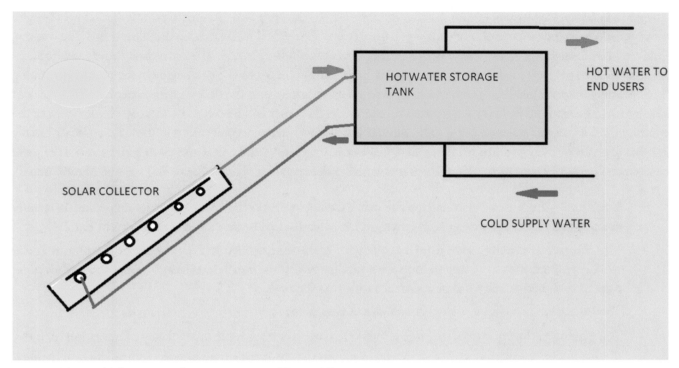

Fig.2.3 Typical Thermosyphon Hot Water Heater System.

Solar Collector for Domestic Hot-Water Systems. There are two types of solar collectors that are utilized in solar thermal domestic hot-water systems which are flat-plate collectors and evacuated tube collectors and the design and construction of these two types of collectors are significantly different. The flat plate collector is usually constructed with an enclosure within which an absorber plate, flow tubes and require insulation are located, supported by necessary inlet and outlet valves along with a working fluid which works inside the flow tubes. The enclosure usually consists of the support structure, bottom and siding and glazing at the top to facilitate the entrance of the solar radiation. The surface of the absorber plate is usually covered with a special coating that has been designed to enhance the ability of the absorber plate to absorb the optimum amount of solar radiation possible from the incident solar radiation and this enhanced absorber plate creates a close approximation of a theoretical black body. According to Planck (1901) "A blackbody is a surface that

- Completely absorbs all incident radiation
- Emits radiation at the maximum possible monochromatic intensity in all directions and at all wavelengths."

These blackbody principles will ensure that the collector absorb the optimum amount of solar radiation possible and transfer the maximum amount of heat energy possible to the working fluid inside the collector flow tubes. The working fluid circulate through the flow tubes which are attached to the absorber plate, absorbing the heat transferred from the absorber plate to the flow tubes and transporting this heat to the point of use or to a storage system where the working fluid transfers the absorbed heat to the incoming cold water, depending on the design of the water heater. In a closed loop system, the working fluid is used to transfer heat to the fluid that will be utilized and in open loop systems the working fluid is the water heated in the collector and it will flow directly to point of use. The solar collector usually consists of an enclosure, the absorber, flow tubes, insulation, inlet and outlet points and glazing cover and all of these components perform a specific function which is essential to the operation of the collector and the water heater overall.

- Enclosure – Acts as a container for the other elements, is usually rectangular in shape and is usually constructed from high strength, lightweight metal such as Aluminum and is as shown in Fig 2.4.

- The absorber is usually made from a metal with high conductivity such as copper or aluminum and is usually painted with a coating to improve its ability to absorb the solar radiation. The absorber flow pipes may be designed in the form of meander or harp as shown in Fig 2.5.

- The insulation is added to reduce losses due to conduction

- The glazing performs several major functions that are very important to efficiency of operation inclusive of minimizing convective and radiant heat loss from the absorber, enhancing the transmission of incident solar radiation to the absorber with minimum loss and to protect the absorber plate from the outside environment (Bakari et al. 2014).

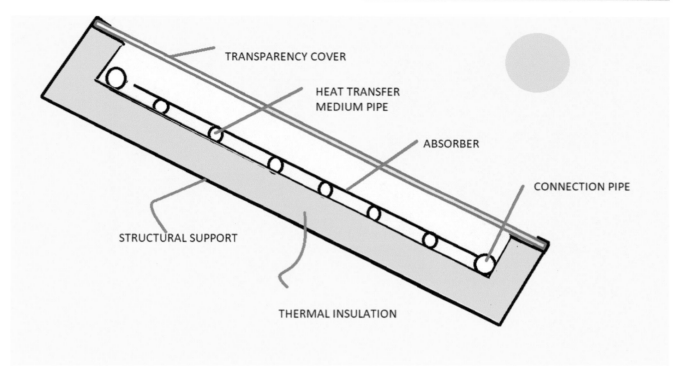

TRANSPARENCY COVER

HEAT TRANSFER
MEDIUM PIPE

ABSORBER

CONNECTION PIPE

STRUCTURAL SUPPORT

THERMAL INSULATION

Fig. 2.4 Typical Flat Plate Solar Collector Section.

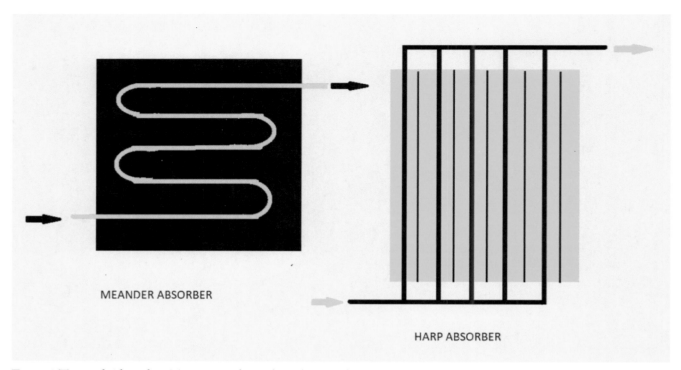

MEANDER ABSORBER

HARP ABSORBER

Fig 2.5 Typical Absorber Types Used in Flat Plate Collectors.

The Evacuated Tube Solar Collector. The evacuated tube solar collector operates on the same principles as the flat plate collector with the exception that the solar energy is collected in double walled glass tubes from which the air has been evacuated to eliminate the insulating effects of air and thereby improving the collection efficiency of the collector.

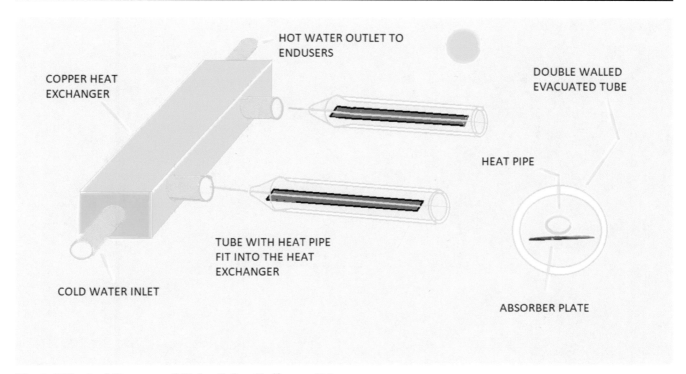

Fig 2.6 Typical Evacuated Tube Solar Collector Diagram.

In general, the capacity of the solar thermal hot water heating systems can be increase by adding collectors and storage tanks or using a larger storage tank and the collectors can be added in parallel or in series depending on the desired outcome, high volume of hot water or high temperature water.

Solar Thermal Systems for Power Generation

The solar thermal systems used for power generation usually fall into a class of solar energy systems referred to as Concentrated Solar Power (CSP) and the Solar Chimney, with the CSPs consisting of a group of four technologies and two systems, Parabolic Trough and Fresnel Concentrator, Line Focus Systems, Dish Sterling and Solar Tower, Point Focus Systems. These CSP systems are utilized to produce high temperature, high pressure fluids that will provide heat and pressure energy that can be converted to mechanical energy in turbines and engines which then uses this mechanical energy as a prime mover to drive a turbogenerator to create electrical energy. Typical systems are as shown in the diagrams below.

Parabolic Troughs

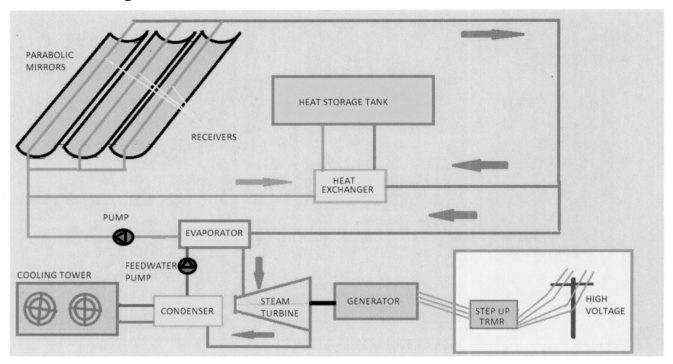

Fig. 2.7 Typical Parabolic Trough Line Focus Electrical Power Generation Plant.

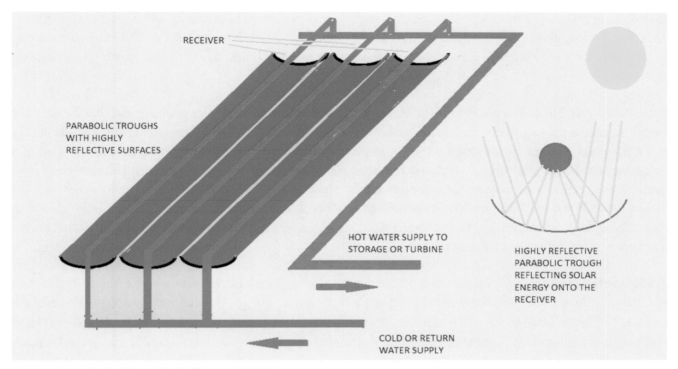

RECEIVER

PARABOLIC TROUGHS
WITH HIGHLY
REFLECTIVE SURFACES

HOT WATER SUPPLY TO
STORAGE OR TURBINE

HIGHLY REFLECTIVE
PARABOLIC TROUGH
REFLECTING SOLAR
ENERGY ONTO THE
RECEIVER

COLD OR RETURN
WATER SUPPLY

Fig 2.8 Parabolic Trough Collector (PTC).

Components of a Parabolic Trough Collector(PTC). The basic components of the parabolic trough collector per IT Power India (2015) are as follows:

- Parabolic Trough Mirrors/Reflectors
- Linear Receiver Tubes
- Trough Stands
- Sun Tracking System
- Heat Transfer Fluid

Parabolic Trough Mirrors or Reflectors. The parabolic trough mirrors or reflectors are usually made from glass or aluminum and they are usually 3-4 mm thick with a parabolic shape, made from tempered and toughened solar grade glass which are tested for hardness and durability. The glass is also provided with a reflective silver back coating with a specific reflectivity greater than 93% and protective edge sealing on all sides of the mirrors. The reflectors may also be 0.3-0.8 mm thick, parabolic shaped anodized aluminum with a PVD reflective coating to provide a specific reflectivity greater than 88% and solar lacquer/Teflon coating/epoxy coating for corrosion protection.

Linear Receiver Tube. The receiver of the Parabolic Trough collector is usually placed at the line focus of the trough to ensure the capture of the focused concentrated solar radiation and the transfer of the thermal energy to the thermal fluid used in the system. The receiver is usually constructed from a metallic tube surrounded by a glass tube and the metallic tube is usually 1-2 mm thick, 304 grade stainless steel coated with a black chrome solar grade absorber paint or selective coating that provides an absorptivity of 0.9-0.95 and an emissivity of $0.09 - 0.015$ and a tube diameter of 25-35 mm. The outer glass tube is usually made from 2-3 mm thick borosilicate glass with a transmissivity greater than 0.95% and a tube diameter of 50-80 mm. The receiver is usually supported on the framework of the mirrors or reflectors.

Trough Stand. The parabolic mirrors or reflectors are usually supported by steel frameworks which are designed to firmly secure, support and transport the mirrors as they move through the different position while they are tracking the movements of the sun on a daily basis. The stands are usually constructed from shaped structural steel which are designed to support the weight of the mirrors or reflectors and wind load specific to the location. The stands must also be protected against corrosion by galvanization or the use of protective coating or paint.

Tracking System. To optimize the operation of the system the mirrors or reflectors must be able to track the movement of the sun across the sky and this can only be achieved by the use of an automatic sun tracking system. The tracking system is designed to enable the parabolic troughs to remain focused towards the sun during daylight hours and thereby ensuring the capture of the maximum amount of solar radiation possible and this tracking system usually consists of the following components electrical motors, gear box, gear and pinion, shaft, solar radiation sensor, sun position sensor, anemometer, feedback mechanism, wind sensor, timer and or a microprocessor and servomotors. The usual system is single axis tracking that moves in an East-West direction.

Heat Transfer Fluid. The heat transfer fluids (HTF) are used as the medium to capture the heat generated by solar radiation in the collectors and this same heat may be transferred to a heat exchanger to generate high temperature, high pressure steam that will drive the turbogenerators directly to generate electrical power or this heat may be transferred to a Thermal Energy Storage facility from where it may be recovered at later time to generate electrical power. According to Brosseau et al (2004) "The current baseline design of SEGS use Therminol VP-1 heat transfer fluid in the collector field." Therminol VP -1 is a thermal oil that has a low freezing point, 12^0 C, and is stable up to 400^0 C which provides the power plants with the ability to use higher pressures and temperatures steam that can be utilized in more efficient turbogenerator sets. The main disadvantages of Therminol VP-1 are that it is relatively expensive and it is not environmentally friendly. The other heat transfer fluid that is currently being used or is currently on the market is molten salts that are usually utilized on the secondary sides in some electrical power generating plants. These salts are usually heated by the Therminol VP-1 in a heat exchanger after which the molten salt is placed in storage or is used immediately to generate steam for the power plant. According to Brosseau et al (2004) efforts are currently being made to develop a molten salt for use in the collector field and TES media in a parabolic trough power plant.

Dish Stirling Solar Power Plant

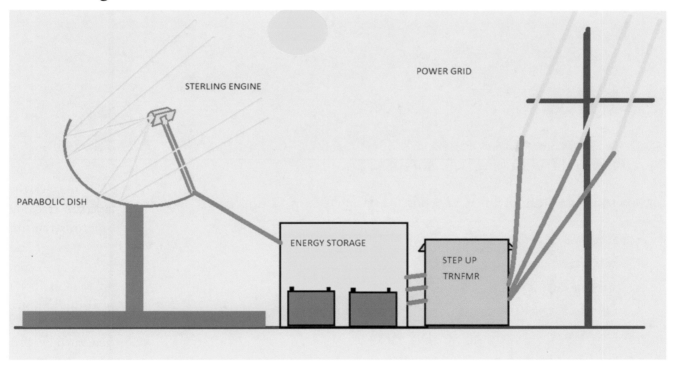

Fig 2.9 Typical Dish Stirling (Point of Focus) Solar Electrical Power Generation Plant

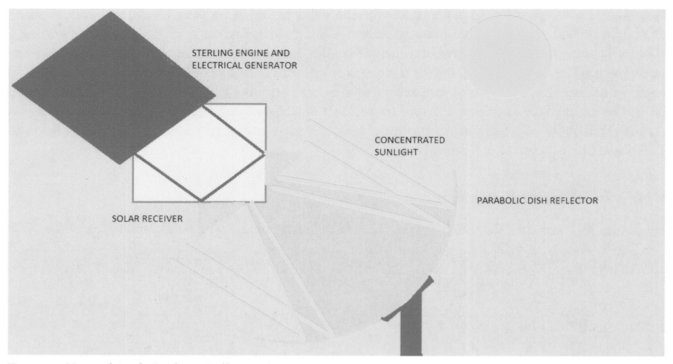

STERLING ENGINE AND
ELECTRICAL GENERATOR

CONCENTRATED
SUNLIGHT

PARABOLIC DISH REFLECTOR

SOLAR RECEIVER

Fig. 2.10 Typical Dish Stirling Collector (DSC).

Components of a Dish Stirling Collector. The components of a Dish Sterling Collector are as follows:

- Parabolic Dish Concentrator
- Solar Receiver
- Sterling Engine
- Electrical Power Generator
- Dish Stand
- Sun Tracking System

Parabolic Dish Concentrator. According to Fraser (2008) The concentrator can be one of three designs:

- A silver/glass parabolic dish with durable mirror surfaces that have reflectivity in the range of 91-95%
- A polymer reflective film that has optical properties of high reflectivity of 94.5%
- Membranes stretched across a hoop or a rim with a secondary membrane placed behind the first after which a partial vacuum then pulls the first membrane into a parabolic shape.

In general, Parabolic Dish Concentrators use a reflective surface of aluminum or silver deposited on glass or plastic, with the silver deposited on glass producing the most durable surfaces. Relatively thin glass, 1mm, is generally usually used to create the curvature required for the short focal lengths of the dishes. Another factor that influences the reflectance produced is the lead content of the glass and it has been found that low iron content glass, thin silvered solar mirrors produce reflectance in the range of 90-94%.

Dish Sterling Solar Receiver. According to the EERE the receiver in the Dish Sterling system is typically a heat exchanger, bank of tubes or heat pipes, that acts as the interface between the collector and the Sterling Engine-Generator set. This heat exchanger absorbs the focused concentrated solar energy and transfers it, in the form of heat, to a heat transfer fluid (HTF) in the Sterling Engine-Generator set where the heat energy is converted to mechanical energy and provides the motive force for driving the generator to create electrical energy. To prevent damage to the receiver due to high flux intensity at the focal point the receiver is usually placed behind the concentrator's focal point with an aperture placed at the focal point to reduce heat losses due to radiation and convection.

Sterling Engine

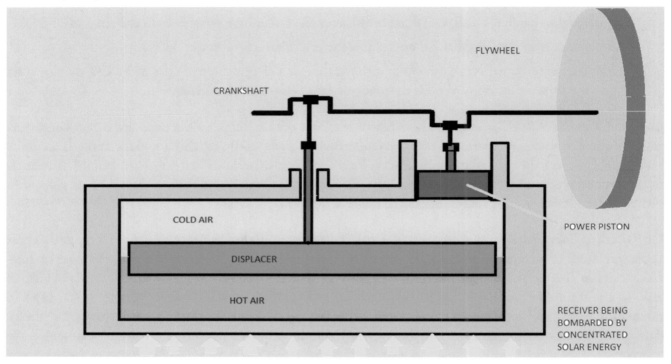

Fig 2.11 Typical Sterling Engine Operations. Source: Nakahara (2008)

The Stirling Engine is a heat engine that uses a source of heat, a fixed amount of a heat transfer fluid (HTF) combined with cylinder, piston and crank shaft to create mechanical energy and motion. The HTF used is normally one of several possible fluids inclusive of liquid sodium, air, helium and hydrogen, fluids that will not be combusted to produce any environmental pollutants or unsafe or unfriendly byproducts. The HTF fluid will never be in direct contact with any flame or source of ignition as it will only be cyclically heated, expanded and cooled to perform the required work and therefore should not physically or chemically degrade, leaking may be the only possible problem. The required work in the Sterling Engine will be done by the pressure difference created by heating and cooling of the HTF in the confined space of the Stirling Engine and the external heat source for providing the required heat is usually a combustible fuel or solar energy.

Electrical Power Generator. The mechanical energy and motion created by the Sterling Engine acts as a prime mover for the electrical generator that is usually attached to the Sterling Engine at the location of the flywheel with the Sterling Engine directly coupled to the generator as shown in Fig 2.11. The spinning of the rotor in the generator across the rotating magnetic fields set up in the stator windings will create electrical energy that may be extracted at the design take of points.

Dish Stand. The parabolic dishes are usually supported by steel frameworks which are designed to firmly secure, support and transport the dishes as they move through the different positions in two planes while they are tracking the movements of the sun on a daily basis. The stands are usually constructed from shaped structural steel which are designed to support the weight of the dishes and wind load specific to the location. The stands must also be protected against corrosion by galvanization or the use of a protective coating or paint.

Sun Tracking System. To optimize the amount of solar energy collected each day by the parabolic dish collectors, it is necessary that the collectors be always directed at the sun and to achieve this they must have the capacity to track the movement of the sun throughout the day. The tracking of the sun requires the utilization of two sets of mechanisms to follow the movements of the sun in two different planes, Z-Y mechanism to track in the horizontal and vertical planes and the X-Y mechanism to track in the left to right plane. The movements in these two planes are usually achieved by using two DC motors, associate gears and accessories and the overall tracking system is made up of a few components inclusive of a microcontroller (Arduino Uno), DC motors (for left-right and up and down movement), DC Motor Driving Cards, Relay Cards, Light Sensor (LDR), gearboxes, linear actuators and resistances (Farsakoglu & Alhamad 2018). The system may also be upgraded by the addition of global positioning system, GPS, windspeed sensors and position sensors (Butti et al 2018).

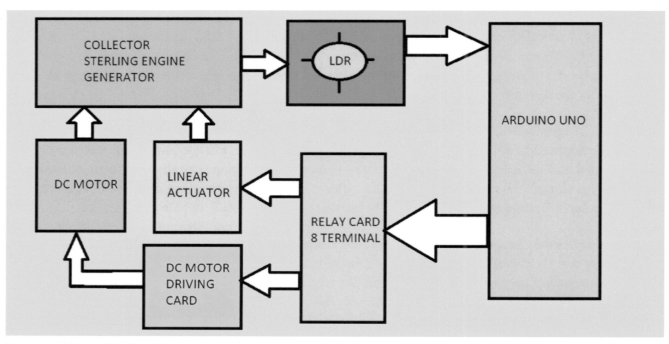

Fig. 2.12 Typical Solar Tracking System Block Diagram.

Linear Fresnel Solar Power Plant

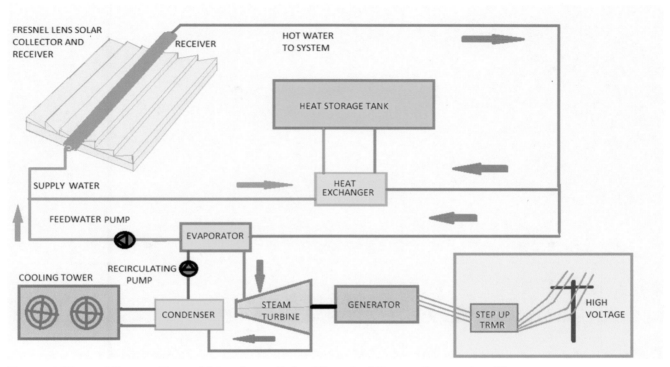

Fig.2.13 Typical Linear Fresnel Line focus Solar Electrical Power Generation Plant.

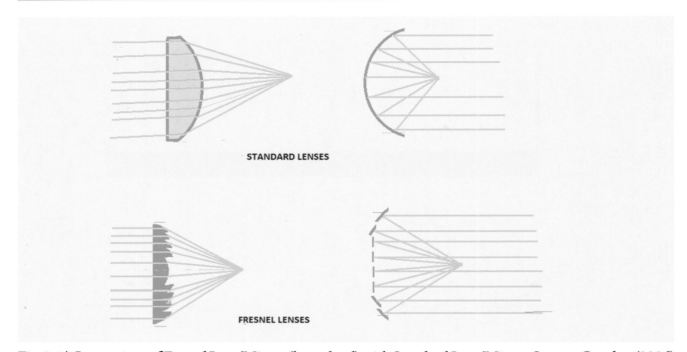

STANDARD LENSES

FRESNEL LENSES

Fig. 2.14 Comparison of Fresnel Lens/Mirror (lower level) with Standard Lens/Mirror. Source: Gunther (2006)

The Fresnel reflector technology was first developed for use in lighthouses by French physicist Augustin-Jean Fresnel in the 18th century and "the principle of this lens is the chopping of the continuous surface of a standard lens into a set of surfaces with discontinuity between them (Gunther 2006). The benefits of the Fresnel design are as follows:

- The reduction in volume and weight of standard lenses.
- The reduction in cost to produce Fresnel lenses
- The reduction in cost to construct support systems for Fresnel systems.
- The ability to produce quality output even with the significant changes in lens design.

The above therefore makes it possible to produce circular and linear Fresnel mirrors and lenses that will work as effectively as parabolic dishes and parabolic troughs without a major reduction in the effective output of the systems that were designed to use parabolic dishes and troughs. The Fresnel lens type power plants provides several advantages relative to parabolic dish and trough power plants (Gunther 2006) and these advantages are as follows:

- Considerably lower investment costs for the solar field at the same aperture area and consequently lower per kilowatt cost for producing the same amount of electricity.

- Lower operation and maintenance cost per kilowatt produced

- Higher land use efficiency.

The only disadvantage is the lower solar-to-electric efficiency.

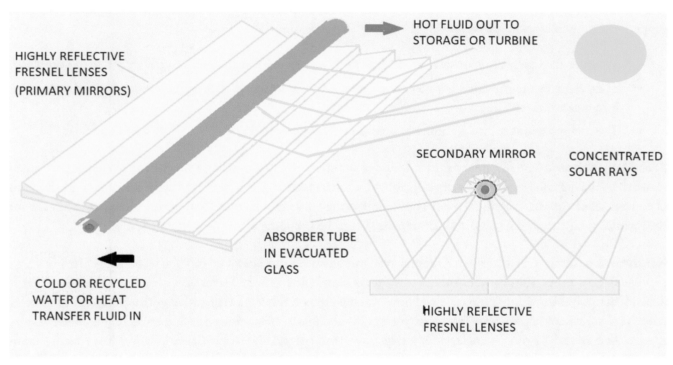

Fig. 2.15 Typical Linear Fresnel Collector (LFC).

Components of the Linear Fresnel Solar Collector

The Linear Fresnel Solar Collector consists of the following components:

- Primary Mirrors
- Secondary Mirror

- Receiver
- Support Stands
- Sun Tracking System
- Heat Transfer Fluid

Primary Mirrors. The primary mirrors used in a Fresnel collector are narrow segments of mirrors or reflectors that are used to create an array of mirrors/reflectors that are all focused to a common focal point and the basic properties of the mirrors are as follows:

- They are elastically curved, solar grade, low iron content mirrors that are scratch resistant and durable.
- These mirrors are provided with a protective and reflective metal coating and weather proof protection for all exposed edges.
- The mirrors and coatings combine to provide a reflectivity of 95%

Secondary Mirrors. The secondary mirror is added to reduce optical losses and the mirror is usually positioned above the absorber tube as shown in Fig 2.15 above and this secondary mirror acts to catch and reflect the fraction of concentrated light coming from the first mirror that may have missed the absorber tube. The material used for the secondary mirror and its required parameters should be the same as the primary mirror.

Receiver. The receiver of the Linear Fresnel system consists of an absorber tube, a secondary mirror and an evacuated glass tube. The absorber tube is a seamless Grade 304 stainless steel tube that has a thickness that is dependent upon the operating pressure and flow of the plant and this tube is provided with a Black Chrome, solar selective absorber coating with an absorptivity greater than 90% to ensure the optimum collection of incident solar energy. This coated SS tube is enclosed in a larger evacuated glass tube to eliminate heat loss due to convection and a cover consisting of a secondary mirror that is located above the enclosed tube to ensure optimum utilization of the focused concentrated solar energy. The receiver may also be provided with insulation at the back of the mirror to prevent heat losses due to conduction.

Support Stands. The Linear Fresnel mirrors are usually supported by steel frameworks which are designed to firmly secure, support and transport them as they move through the different position while they are tracking the movements of the sun on a daily basis. The stands are usually constructed from shaped structural steel which are designed to support the weight of the mirrors and wind loads specific to the location. The stands must also be protected against corrosion by galvanization or the use of protective coating or paint.

Sun Tracking System. The mirrors of the Linear Fresnel Plant are provided with the capacity to track the path of the sun to ensure that the maximum amount of incident solar radiation possible is captured each day. The mirrors are provided with a single axis tracking system that either moves north-south or east -west and this system has the following components:

- DC Stepper motors
- Gearbox, chain and sprocket and support rollers
- Solar Radiation Sensors
- Sun Position Sensors
- Wind Sensor
- Microcontroller with requisite software and GPS algorithm.

Heat Transfer Fluid. Heat transfer fluids (HTF) are used to capture the heat generated by solar radiation in the collectors and the heat generated is usually transferred directly to a heat exchanger where it generates high temperature and high-pressure steam from where it is directed to the power plant to generate electrical power or it may be directed to a Thermal Energy Storage facility from where it may be recovered for use at a later time to generate electrical power. According to Brosseau et al (2004) "The current baseline design of SEGS use Therminol VP-1 heat transfer fluid in the collector field." Therminol VP -1 is a thermal oil that has a low freezing point, 12^0C, and is stable up to 400^0C which allows the power plants to use higher pressure and temperature and more efficient turbines. The disadvantages of Therminol VP-1 are that it is relatively expensive and it is not environmentally friendly. The other heat transfer fluid that is currently used in some solar plants is molten salts that are usually utilized on the secondary sides in power plants. These salts are usually heated by the Therminol VP-1 in a heat exchanged after which the molten salt is placed in storage or used to directly generate steam for the power plant.

Solar Tower Power Plant

The solar tower system is a combination of a field of solar mirrors which collects and concentrates solar radiation and a receiver located high on a tower to produce high temperature and pressure fluids that can be utilized in combination with heat exchangers and a turbogenerator set to produce electrical power. The main components of the solar tower systems are the heliostat array or field, the tower and receiver, the sun tracking system, thermal energy storage units, steam systems, heat exchangers, cooling towers and accessories, turbogenerator sets and the heat transfer fluid (HTF). In this CSP a large array of heliostats(mirrors) collects, concentrate and reflect solar radiation onto a receiver, heat exchanger, that is located on a high tower where the concentrated solar radiation is converted to heat energy in the receiver and this heat energy is subsequently absorbed by a heat transfer fluid

(HTF) that cycles continuously through the receiver. The absorbed heat energy can then be routed along one of two paths, one to a thermal energy storage facility from which it can be withdrawn and utilized at a later time and the other path directly to a heat exchanger where working fluid is used to generate high temperature and high pressure steam that will be used to drive a turbogenerator set that is used to generate electrical power. These solar towers in combination with the appropriate HTF can achieve temperatures in the order of 1000^{0} C which has the capacity to significantly improve power cycle efficiencies.

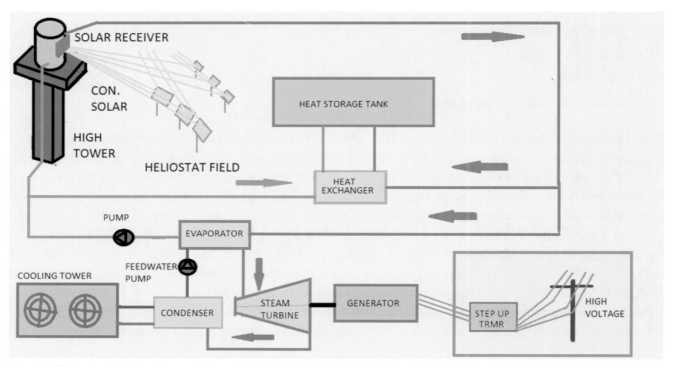

Fig. 2.16 Typical Solar Tower Electrical Power Generating Plant.

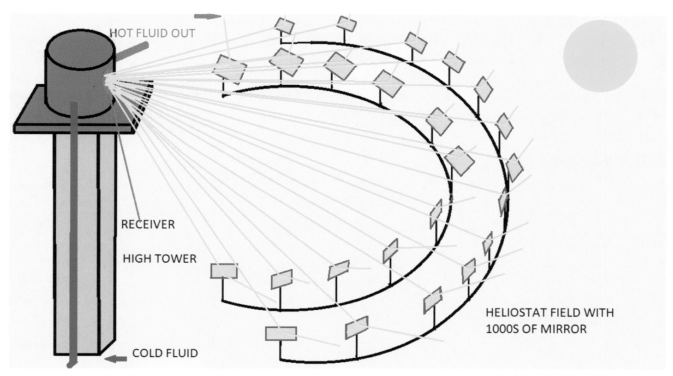

Fig. 2.17 Typical Solar Tower Collector.

Solar Tower Collector (STC). The solar tower collector consists of two main elements, the Heliostat Array or Field and the receiver that is located on the high tower and the heliostats are supported by several systems inclusive of the stands or racks, sun tracking system, controls, computer and software, mechanical and electrical systems.

Heliostats Field. By general definition a heliostat is a computer-controlled mirror or reflector which keeps solar radiation focused on a target as the sun moves across the sky and the components of this heliostat are the mirror module, drive mechanisms, control systems inclusive of position sensor, interface with power system and heliostat field controller, drive controller master control interface electronics for local control of heliostats, computers, software and timers. The major component are the special mirrors that are nearly flat but which has some curvature to focus the Sun's image on the receiver in the tower that can be as much as 100-1000 meter away from it. The mirrors used in heliostats are usually made from thin silvered glass, low iron content preferable, aluminum or silvered stretch polymer membrane, with the glass mirror, with high specular reflectance, 91%, high durability, modest reduction in reflectivity over concentrator lifetime with a long life being the initial choice of industry.

However, silvered stretched polymer membranes are gaining high acceptance due to their lower weight, high flexibility and lower first cost, which when combined reduces the cost of necessary structural supports, tracking systems and the overall operations cost (Jorgensen et al 1994 NREL).

To be effective and to provide the level of concentration necessary, heliostats are usually assembled in large arrays or fields with the size of the field depending on the level of power output to be generated.

Solar Chimney

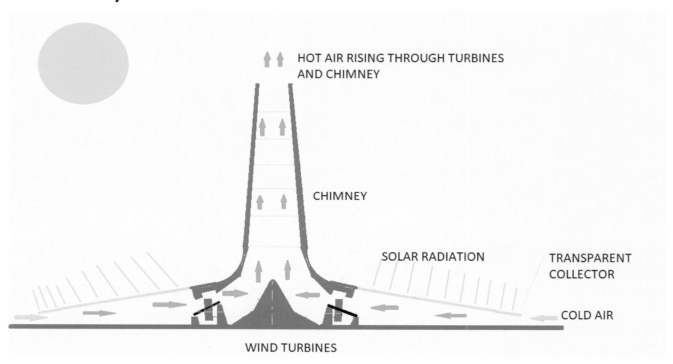

Fig. 2.18 Typical Solar Chimney Collector (SCC).

The solar chimney is a fairly simple machine that is based upon the principle of convection currents, hot air rising and heavier cold air falling, and this principle is combined with three components that make up the solar chimney. These three components are collector, wind turbines and chimney or tower.

Collector. The collector is made up of transparent glass or plastic that is attached to a support structure above ground level that allows solar radiation to pass through to the ground approximately two meters below the collector and the incident solar radiation heats up the ground directly beneath the collector. The heated ground then heats up the surrounding air, changing its temperature and density, which causes the heated air to rise and move towards the chimney, which is at the center of the structure, where it passes through the wind turbines, which are located in the base of the chimney, and then on up into the tower, before exiting at the top of the tower. The collector is usually constructed surrounding the base of tower and it is usually inclined at an angle to allow the smooth upward flow of the hot air towards the turbines and tower. The collector is usually supported in a manner to optimize the area exposed to the direct rays of the sun and the movement of the heated air.

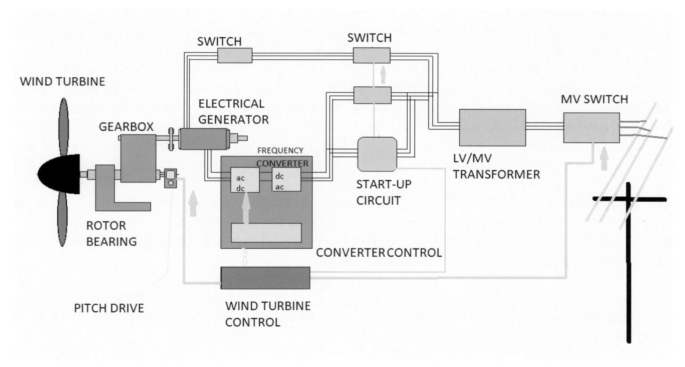

Fig. 2.19 Typical Wind Turbine Electrical Power Generation Plant Utilized In A Solar Chimney.

Turbine. The turbines are usually special design turbines, as depicted in Fig. 2.19, and are usually driven by the pressure difference created across its blades by the hot air and the chimney, with the height of the chimney being a determining factor of the velocity developed by the turbines and the amount of electricity generated and supplied to the grid or other take-off points. The wind turbine power generation plant is usually made up of several major components inclusive of:

- The blades, support bearing and housing,
- The gear box to achieve the required revolutions per minute to match the generator,
- The generator which provides all the components required to create the electricity inclusive of fixed stator winding and require rotating core, bearings, and necessary generator control systems.
- Turbine controls which include speed controls, converter controls, necessary electrical circuits and transformers.

Tower. The chimney or tower is the center piece of this power plant and its main functions are, the creation of the required pressure difference across the turbines, the necessary suction to pull air under the collector, hot air through the turbines and to keep the process going continuously. The height of the tower and the temperature difference created in the collector are both responsible for the air velocity, typically 15m/s, passing through the turbines and the tower and it is this air velocity that imparts kinetic energy to the turbines.

The operating capacity of the Solar Chimney may be increased by placing heat capturing and storing devices on the ground under the collectors, these devices are typically water filled containers that capture the heat of solar radiation during the hot days and releases it at night when temperatures have fallen and allows the chimney to operate much long than the hours when direct solar radiation is available.

The electrical output from the Solar Chimney may also be increased by creating a Hybrid Solar Chimney by adding solar cells to the external surface of the tower or by using BIPV material to form the external skin of the tower.

CHAPTER 3

Solar Photovoltaic Energy Systems

Photovoltaic energy systems are based upon the phenomenon that solar radiation causes electrical energy to flow in certain materials mostly semiconductors, inclusive of silicon, selenium and an assortment of other exotic material inclusive of copper, gallium, cadmium, germanium, indium and alloys of several metals, with selenium being the first material tested and proven to demonstrates this phenomenon and selenium was later followed by silicon which currently dominates the solar PV cell market. According to the Merriam Webster definition, photovoltaic is "relating to the generation of a voltage when radiant energy fall on the boundary between two dissimilar substances". What is today known as the Photovoltaic Effect was first observed by a young French scientist in 1839, Edmond Becquerel, who had been experimenting with an electrolyte and two metal electrodes, he observed that when exposed to light the cell would emit electrical energy. The growth in understanding and utilizing this effect occurred in many steps over an extensive period of time, 1839- present, with inputs from many different scientists and organizations inclusive of:

- Willoughby Smith discovers the photoconductivity of selenium between 1873-1876
- The publication of a book, The Action of Light on Selenium, in 1877 by two American scientist, William Adams and Richard Day.
- The development of the first photovoltaic (PV) cell by Charles Fritts, an American inventor in 1883. This first cell consisted of selenium attached to a metal plate.
- Heinrich Hertz first observes the photoelectric effect, the phenomena of light freeing electrons from a solid surface to create electricity, in 1887
- The US Patenting of the first solar cell in 1888 by Edward Weston, an American chemist.

- The development of the first solar cell using the outer photoelectric effect, 1888-1981, by Alexander Stoletov.
- Seminal works, Methods of Utilizing Radiant Energy and Apparatus for Utilization of Radiant Energy, by Nikola Tesla earns him recognition and two patents in 1901.
- The development of the semiconductor-junction solar cell by Wilhelm Hallwachs in 1904.
- The development and presentation of his theory of "Photoelectric Effect" by Albert Einstein in 1905.
- Proving Einstein's "Photoelectric Theory" was completed by Robert Millikan in 1916.
- The development of Cadmium Selenide as photovoltaic material by Stora and Audobert in 1932.
- Bell Laboratories display a new high powered solar cell that has an efficiency of 6% in converting solar energy to electrical energy in 1954.
- Silicon Solar cells are produced commercially by 1953-1956
- Western Electrics begin the commercialization of silicon based photovoltaic cell systems design technologies in 1955.
- The first solar-propelled space satellite, US Vanguard 1, was launched in 1958.
- The development of a radiation resistant solar cell by the US Signal Corps Laboratories in 1958
- Hoffman Electronics develops a solar cell that is 9% efficient in 1958 and a 14% efficient one in 1960.
- Sharps Corporation develops a workable module of silicon solar cells in 1963
- Japan installs 242 watts PV array on lighthouse in 1963
- NASA orbits Astronomical Observatory using PV array in 1966
- The Soviets introduce the first solar powered manned space vehicle, the Soyuz 1, in 1967
- To foster and enhance progress with photovoltaics and solar energy Japan launches a program call "Project Sunshine" in 1974
- The US Department of Energy founds the US Solar Research Institute in Golden Colorado in 1977.
- The Institute of Energy Conversion at the Delaware University develops the first thin film solar cell, with an efficiency of 10%, in 1980.
- A 20% efficient silicon cell is developed by the Center for Photovoltaic Engineering in 1985.
- Reflective solar concentrators are first used with solar cells in 1989.
- The development of the first efficient photo electrochemical solar and dye-sensitized cells are developed in 1991.

- University of Florida creates a 15.89% efficient thin film cell in 1992.
- NREL Labs develop new solar cells from gallium indium phosphide and gallium arsenide that develops efficiencies in the region of 30% and thin film solar cell that reach 32 % by 1999.
- The inverted metamorphic triple-junction solar cells are developed in 2008
- Flexible printed solar panels are on the market in 2015
- Sunless solar power is discovered in 2016.

Photovoltaic (Solar) Cell

Photovoltaic (PV) cell technology is based upon semiconductors and according to Segal (2021) "A semiconductor is a material product usually comprised of silicon, which conducts electricity more than an insulator, such as glass, but less than a pure conductor such as copper or aluminum." Semiconductors are mostly made from silicon crystals, mono and polycrystalline, and in lesser volumes from other materials such as germanium and gallium arsenide, and it is usually treated or doped with different substances to create two basic types of semiconductors, the Negative Type (N-Type) and the Positive Type (P-Type). To create the N-Type semiconductor the silicon is usually treated with one of the following, Antimony, Arsenic or Phosphorus, while to create the P-Type semiconductor the silicon is treated with one of the following material Boron, Aluminum or gallium. The doping creates the necessary condition in each type of semiconductor to allow them to function as required, in the N-Type it creates an excess of electrons which are negatively charged and with the P-Type there is a paucity of electron which creates holes that behave as positive charges.

The PV Cell is made up of two or more layers of semiconductors, one N-Type layer and one P-Type layer which are bonded together and when exposed to sunlight some photons of light are absorbed by the N-Type semiconductor atoms and these photons of light liberate electrons from the N-Type layer subsequent to which these electrons flow through an external circuit back into the P-Type semiconductor causing the production of an electric current.

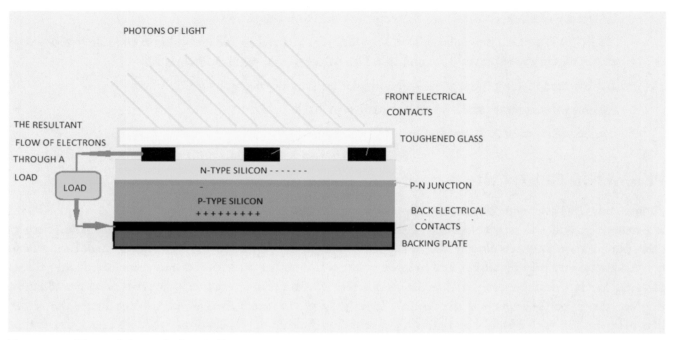

Figure 3.1 Typical Basic Solar Cell Construction.

There are currently two types of Photovoltaic cells, wafer and thin film, both of which are currently used in the PV industry, with the wafer type currently used in 90% of solar energy systems due to its ability to convert photons of light to direct current at a much higher level of efficiency than the thin film type. The thin film type is however less costly to produce and will facilitate greater penetration of solar energy systems into the energy market, thereby making it more accessible to poorer nations and people.

Wafer Type Photovoltaic Cells

Photovoltaic cells are relatively small 5"x5" or 6"x 6" and 200-300 microns thick and produce a small voltage 0.5 – 0.6 volts per cell, depending on the material that the cell is made from, with silicon cells producing 0.5Volt per cell, and to increase their capacity to meet domestic, commercial and utility demands multiple cells must be wired together in series and or parallel to increase both voltage and current output. The standard combination of cells is called modules, panels and arrays, with each module containing 36 cells with the capacity to produce 18 volts, each panel containing 3 modules with the capacity to produce 54 volts and arrays containing as many panels or groups of panels as are required to generate the required amount of power, voltage and current.

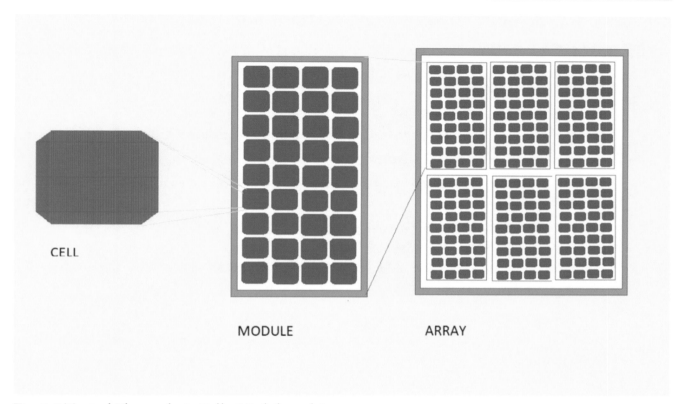

CELL

MODULE

ARRAY

Fig. 3.2 Typical Photovoltaic Cells, Module and Array.

Typically, a complete module ready for installation will consist of the following components, an aluminum frame as structural support, a sheet of tempered glass to provide necessary protection from the many elements of nature while ensuring the module will be continuously exposed to sunlight for as many hours as there are in a day, a top and bottom encapsulant sheets to maintain the integrity of the solar cells in the module, the module, which fits between the top and bottom encapsulant sheets, a backing sheet and a junction box which is used to connect module to necessary electrical systems.

The Manufacturing of PV Crystalline Silicon Wafers

The typical PV Cell is made from silicon which has been produced using several processes, a few of which are:

- The mining of the silica or quartz.

- The processing of these ores to produce elemental silicon.

- The further processing of the silicon to eliminate all impurities to produce 99.9999 percent pure silicon.

- The further processing of the pure silicon to produce monocrystalline silicon.

- The processing of the pure silicon scrap and other polycrystalline material to produce to produce polycrystalline silicon.

- The processing of this pure silicon to create doped monocrystalline and polycrystalline silicon that are produced in cylindrical and square ingots respectively.

- The slicing of both monocrystalline and polycrystalline ingots into 200-300 microns thick wafers

- The polishing of these wafers to obtain smooth surfaces on both sides of the wafers.

- The assembling of semiconductor into cells, modules and arrays.

- The necessary encapsulation and provision of support structures to make them ready for producing electrical power.

An examination of the steps involved in producing solar cells this way will reveal significant areas of waste, inclusive of the silicon wasted when the ingots are sawed into thin slices, the polishing of the wafers to produce smooth surfaces and the required trimming of the monocrystalline cylindrical ingot crystals into a true cylinder. When added to the required thickness of the wafers, 100 microns, minimum, as thinner wafers would require the use of sophisticated light trapping systems to produce at the optimum level, all of these material losses add up to make the production of these semiconductors relatively expensive and makes the cost to produce electricity from them relatively high.

The steps in making pure monocrystalline silicon are as follows:

- To make silicon on an industrial scale silica, sand or quartz, is processed by heating it to approximately 2000 ^{0}C and combining it with carbon to produce impure low purity, 97%, silicon (Ball, 2003).

- This low purity silicon is further processed in several steps to produce very high purity silicon, 99.9999% pure, these steps include, grinding the 97% pure silicon to a powder, heating it to 300^0 C in the presence of gaseous hydrogen chloride to produce trichlorosilane. This trichlorosilane is then vaporized at 1100^0 C in the presence of hydrogen gas to produce 99.9999% pure silicon. The silicon made in this way only produces polycrystalline crystals which are not the best for making semiconductors and must be further processed to produce monocrystalline silicon, the gold standard for the industry.

- Monocrystalline silicon for semiconductor is made by utilizing the Czochralski Process which consist of the following steps, melting and doping the polycrystalline silicon in a container, introducing a pure monocrystalline seed into the melted mix, initialization of the crystal growth process and pulling the crystal, which forms an imperfect cylinder of pure monocrystalline silicon.

PV solar systems made with crystalline silicon has overtime demonstrated qualities that make the demand for them still relatively high, 95 % of all PV systems on the market are made from crystalline silicon, even in the face of new advancing technology. Some of these qualities include:

- Durability. These crystalline cells have great stability and reliability as they can operate in all outdoor conditions without any degradation in their capacity to produce for 20-30 years.
- Relatively high efficiencies. According the Vonderhaar (2017) Monocrystalline and polycrystalline silicon systems have efficiencies of 25 and 20 % respectively in the laboratory and 15-18% in commercial systems, with monocrystalline cells being 5% more efficient than polycrystalline cells. It is also projected that with the addition of new technologies such as sunlight concentrators and multijunction cells the monocrystalline cells could reach efficiencies as high 86%, Green and Bremmer (2016), Vonderhaar (2017).

The main disadvantages of monocrystalline silicon are the high cost to produce 99.999 pure silicon, the higher cost to produce monocrystalline silicon and the significant waste that occurs when shaping and cutting the high-cost pure silicon into thin wafers that are the basis of solar cells.

Thin Film PV Solar

Thin film PV solar systems are an outgrowth of the original PV solar systems that were designed and built to produce electrical power for consumption in all spheres of human activities, however, thin film PV Solar system was also designed to deliver this power at a much lower cost to the consumer. According to Wasa, Kitabatake and Adechi (2004) and Jameel (2015) "a thin film is a small dimensional material on a substrate produced by intensifying one-by-one and ionic/molecular/atomic species of matter" and the thickness of the thin film is usually less than 10 microns. The creation of a PV systems could therefore transition into using 10 microns or less of material to produce the same photovoltaic effect as originally obtained when using 150 mm by 200-300 microns thick material.

The production of thin film PV systems is relatively inexpensive and requires much fewer steps when compared to the production of crystalline solar cells and according to Minneart the competitiveness and desirability of a PV module is highly dependent on the cost to produce each unit of power and thin film cells have the capacity to produce power at a relatively lower per unit cost. Thin film cells can be manufactured in significantly different ways from the typical crystalline silicon cell to great economic advantage, and these ways include:

- Utilization of less PV material
- Utilization of several different materials thereby eliminating total dependence upon silicon
- Simplified manufacturing with fewer processing steps
- Reduced material handling
- Greater capacity to automate manufacturing process
- Integrated monolithic circuit design and the elimination of the individual assembly of solar cells

The divergence in the process of manufacturing of silicon solar cells versus thin film cells is a direct function of several new materials and the modification of silicon that were developed over time that are now available to industry, with several other material still at the experimental stage. The new materials currently available includes:

- Amorphous Silicon (a-Si) – Currently in use
- Cadmium Telluride (CdTe) – Currently in use
- Copper Indium Gallium Di-selenide/sulfide (CIGS) - Currently in use
- Thin film silicon – Currently in use

Thin Film Manufacturing Processes

Over time several methods of depositing thin film on a substrate have been developed and applied in many different areas of industry and most of the methods developed can be classified under two categories, Pressure Vapor Deposition (PVD) and Chemical Vapor Deposition (CVD). The PVD category contains two methods, Thermal Evaporation and Sputtering while the CVD category contains several methods inclusive of Plasma

Enhanced Chemical Vapor Deposition (PECVD), Laser Chemical Vapor Deposition (LCVD) and Metal Organic Chemical Vapor Deposit (MOCVD). The PVD methods are highlighted in Fig. 3.3 and 3.4. and one CVD method is highlighted in Fig. 3.5

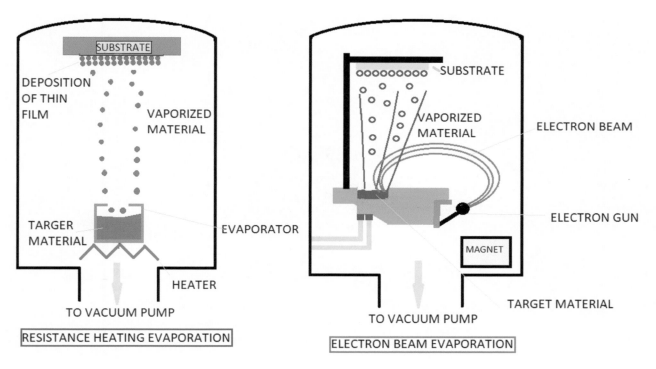

Fig. 3.3 Typical PVD Thermal Evaporation Methods of Forming Thin Film.

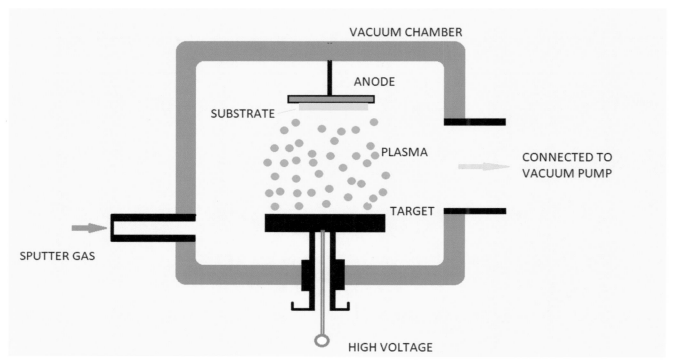

Fig. 3.4 Typical PVD Sputtering Vacuum Deposition Method of Forming Thin Film.

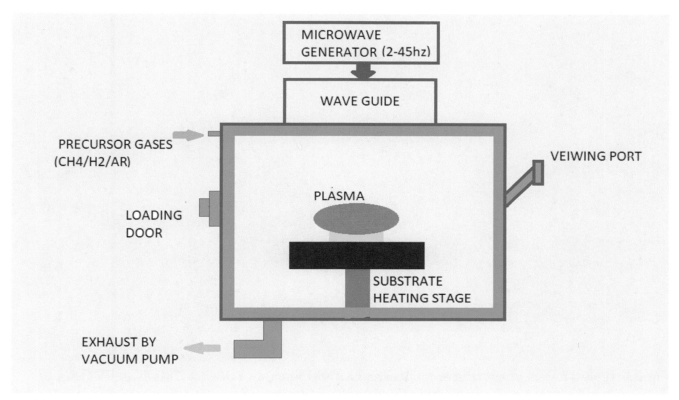

Fig. 3.5 Typical CVD Plasma Enhanced Chemical Vapor Deposition Method of Forming Thin Film.

All of these different methods of creating thin film facilitated the development of the single layer thin film PV material with only one junction, where the important work takes place, and multi layers of PV material creating multiple junctions that significantly increased the activity and productivity of the PV system. Both single layer and multi-layer thin film systems are highlighted in the diagrams below.

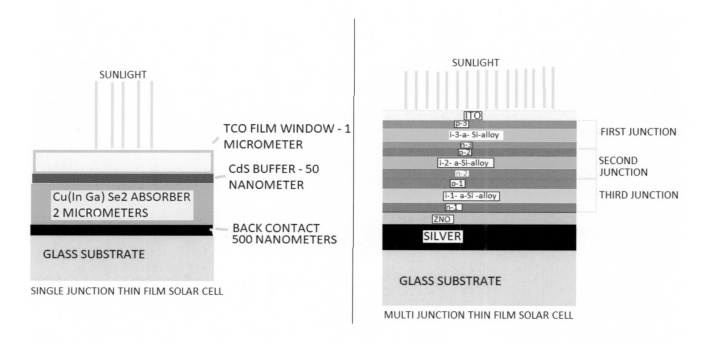

Fig. 3.6 Typical Layers of material on a CIGS and a Multi Junction a-Si-alloy Thin Film PV Cells.

The creation of the multi-junction systems greatly improved the capacity and efficiency of the thin film PV systems and it has been the basis of many systems used in the space industry and in Concentrated Photovoltaic (CPV) systems.

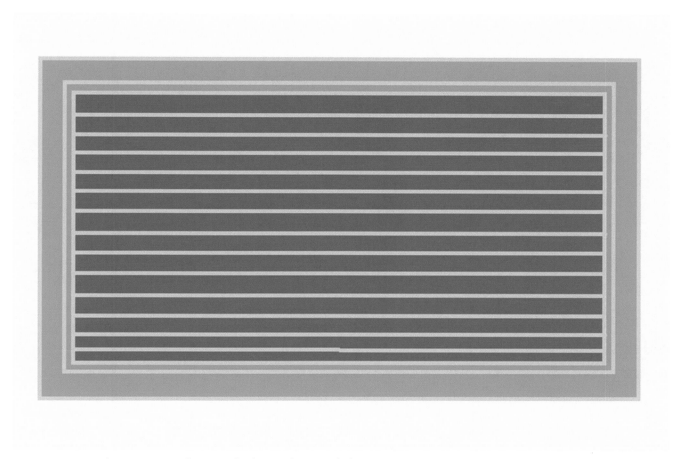

Fig. 3.7 Typical Structure of a Rigid Thin Film Module.

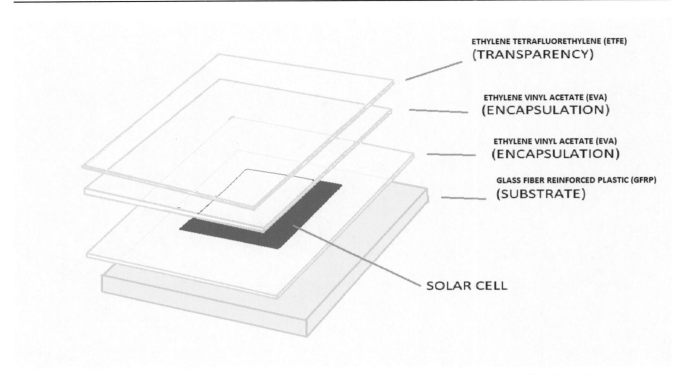

ETHYLENE TETRAFLUORETHYLENE (ETFE)
(TRANSPARENCY)

ETHYLENE VINYL ACETATE (EVA)
(ENCAPSULATION)

ETHYLENE VINYL ACETATE (EVA)
(ENCAPSULATION)

GLASS FIBER REINFORCED PLASTIC (GFRP)
(SUBSTRATE)

SOLAR CELL

Fig. 3.8 Typical Thin Film PV on Flexible Substrate. Source: Fraunhofer Institute for Solar Energy

Thin film PV systems can be provided in two basic formats, rigid with glass substrates as shown in Figs 3.6 above and as flexible form with fiber reinforced plastic substrate as shown below in Fig. 3.8.

All of the systems described above for the application of different layers of material to create the thin film PV can be automated on a production line to produce thin film solar modules in large volumes at costs that are much lower than that required to produce solar cells from crystalline silicon. This reduction in cost to produce thin film PV is significant, however, thin film PV is less efficient than crystalline PV solar cells, with the efficiency of thin film PV in the range of 7.6 - 12.1% based on the material of construction, with CIGS at the highest level and a-Si at the lowest level, while crystalline silicon solar PV has commercial efficiency in the region 15-18%.

Company	Material	Area (cm²)	Efficiency (%)	Power (W)
United Solar	a-Si Triple	9276	7.6	70.8
Solar Cells Inc.	CdTe	6728	9.1	61.3
Solarex	a-Si dual junction	7417	7.6	56
APS	a-Si/a-Si	11522	4.6	53
Siemens Solar	CIS	3664	11.1	40.6
BP Solar	CdTe	4540	8.4	38.2
United Solar	a-Si Triple	4519	7.9	35.7
Golden Photon	CdTe	3366	9.2	31
ECD	a-Si/a-Si/a-SiGe	3906	7.8	30.6

Table. 3.1 Typical Thin Film PV Efficiencies. Source: NREL

Wafer PV Versus Thin Film PV

There are several major differences already established between wafer and thin film PV inclusive of the size of the cells, the type and amount of material making up each type of PV system and the different efficiencies as shown above, however, there are several other significant differences that must be noted. According to the Fraunhofer Institute for Solar Energy some of the differences are as shown in the table below.

Conventional Crystalline Silicon PV Panels	Thin Film Panels
Rigid	Flexible
Heavy	Lightweight
Glass	Plastics and polymers
Costly installation	Lower cost installation
Roof penetration	No roof penetration
Unsightly	Low profile
Low power production per panel	High power production per panel
High cost to produce and install each panel	Low cost to produces and install each panel

Table 3.2 Differences Between Wafer and Thin Film PV. Source: Fraunhofer.de

The information in Table 3.2 indicates that thin film panels have significant cost advantages over wafer panels and this cost advantage significantly reduces the cost to produce and install them, thereby making them more accessible to a much larger portion of the global population. This greater availability, will allow for a greater uptake of the solar renewable technology which will displace the demand for electrical power produced from fossil fuels. One of the disadvantages of the thin film PV system however, is that the advantage gained in the reduction in the amount of material used to create each modules can lead to significant problems as the smaller size permit the installation of many more cells into each and this increase number of cells allows thin film modules to produce power at higher voltages and the higher voltage produced per panels places a limiting factor on the use of thin film panels in large utility scale plants as it limits the number of panels that can be attached in series on any one string without the use of a power optimizer.

Concentrated Photovoltaic (CPV)

Standard PV systems require two basic components to create the basic photovoltaic effect, PV cells made from wafers or thin film semiconductors and direct or diffused sunlight, however, the use of concentrated sunlight in PV systems require an additional component a concentrating mirror or lens, similar to those used in CSP systems, which has the effect of producing significant improvements in energy production, conversion efficiency and a significant reduction in the amount of semiconductor material required in each PV system. The use of concentrated sunlight in PV systems comes with some draws backs that require significant design modifications to mitigate against the problems which arise from the use of concentrated sunlight on PV cells, problems which must be minimized to enable the CPV systems to produce at optimum levels. The major problem associated with CPV systems is the heat that is generated by the concentrated sunlight which can cause excessive heating of the semiconductors and if this excessive heat not dealt with it has the capacity to reduce electricity production as semiconductors operate more efficiently at lower temperatures. The semiconductors themselves will eventually suffer much damage due to overheating if actions are not taken to eliminate the problems associated with excessive heat. The solution designed and implemented for the elimination of the overheating problems was utilization of an appropriately sized cooling systems that was designed to effectively transfer the heat away from the semiconductors.

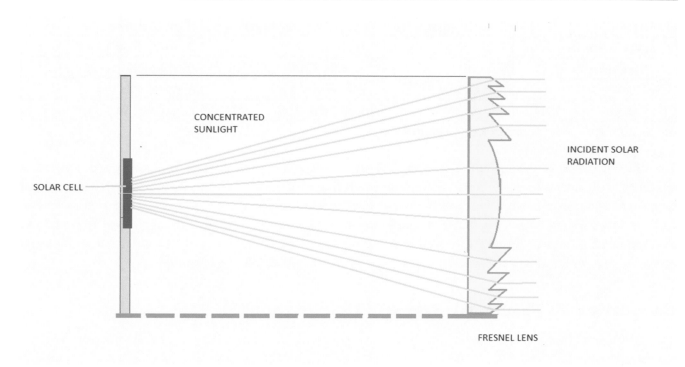

Fig 3.9 Typical Concentrated Photovoltaic System.

The efficiencies achieved by CPV systems are in the region of up to 30%, utilizing a variety of cell materials inclusive of silicon, CdTe, CIGs and an a-silicon alloy as used in multijunction cells, shown in Fig 3.6 above, with the multijunction cells having the highest efficiency of 30 % in field conditions and 40 % in laboratory conditions (Kurtz 2011). There are generally three types of CPV systems, Low-Concentration, Medium-Concentration and High-Concentration, with each level of concentration generating electricity at different levels of output and efficiencies and requiring different levels of cooling and tracking. The different requirements are highlighted in the table below.

ITEM	Low Concentration	Medium Concentration	High Concentration
Concentration Ration	2-10	10-100	100-400 (and above)
PV Material	Silicon	Silicon, CdTe	Multijunction cells
Cooling	Not required	Passive cooling	Active cooling
Tracking	Not required	1-axis tracking	2-axis tracking

Table 3.3 CPV Concentration Level Requirements.

The different levels of concentration provide different levels of electrical energy production and conversion efficiency, with the higher levels of concentrations producing a higher level of output, but these levels of output require the addition of cooling and tracking systems which increases the cost to produce each watt of electricity. As shown in the table above, a system that tracks the movement of the sun across the sky is very important to medium and high concentration PV systems based on the fact that CPV systems can only use direct sunlight and therefore requires the ability to track the direct sunlight for all available hours of sunlight.

Components of a Photovoltaic System

The basic components of a Photovoltaic System are photovoltaic array, charge controller, battery, inverter, maximum power point tracking (MPPT), power optimizer, DC-DC Converter, wiring, tracking and mounting systems, however, some systems do not have tracking systems, charge controllers and batteries.

Photovoltaic Array. The Photovoltaic array is typically the solar collectors for the system which usually consists of an assembly of solar modules to create the desired collector size to meet the power demands of a domestic, commercial or utility user. The array may be located on roofs or on the ground using a variety of mounting systems inclusive of stand-off mounting, racks and poles with or without solar tracking systems. Standard photovoltaic array systems are also complimented by a variety of integrated roofing systems inclusive of the three-tab PV roofing shingle, PV roof slates, PV Atlantis sun slates, solar electric metal roofing, standing seam roofing and other BIPV systems.

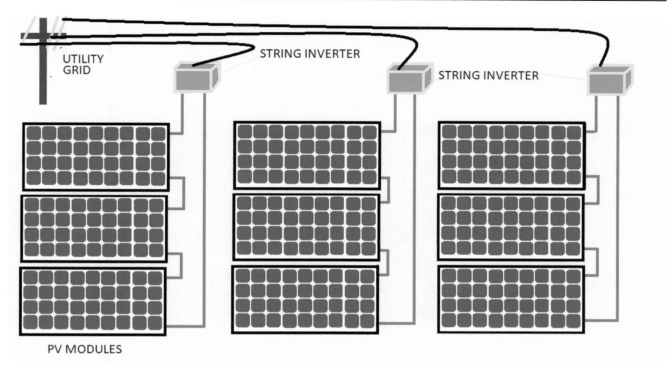

Fig. 3.10 Typical PV System Array (Multi-String Inverters).

Charge Controller. The Charging Controller as designed serves several very important functions for the photovoltaic systems inclusive of the following:

- Battery overcharge and over-discharge protection
- Battery reverse polarity protection at the battery connection terminal
- Battery over temperature protection
- Short circuit protection at the load terminal
- System low voltage disconnection and reconnection
- Optimization of charging efficiency
- The capacity to interface with different types of batteries
- Pulse Width Modulation (PWM) charging with 3-stage charging control which allows the battery to remain unattended for significant periods of time

- Reverse polarity protection at PV panel connection
- PV panel system protection from high battery current

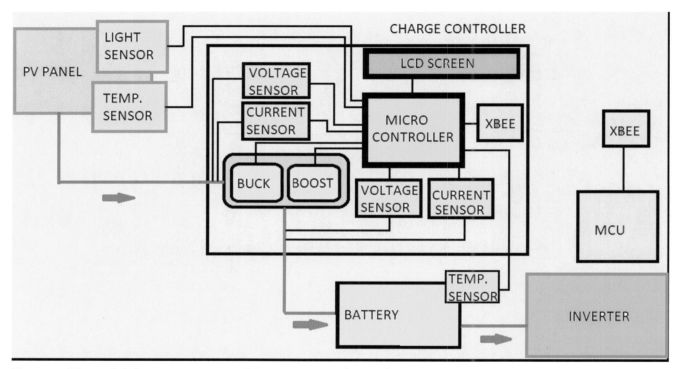

Fig. 3.11 Typical Schematic Layout of Components of PV Charge Controller.

Source: elprocus.com

Batteries. The basic function of batteries in photovoltaic systems according to Dunlop (1997) are

- Energy storage and autonomy – To store excess electrical energy produced by the photovoltaic system and to return this energy when electrical demands require it.
- Voltage and current stabilization – To provide stable voltage and current to electrical loads by suppressing or smoothing out transient that can occur with the generation of electrical power by PV systems
- Supply surge currents – To provide electrical power at peak periods of usage when the demand for electrical power may exceed the production capability of the PV system.

According to IRENA (2019) "Battery storage increases flexibility in power systems enabling optimal use of variable electrical sources like solar Photovoltaic, PV, and wind energy". Batteries designed for use with PV systems while similar to those used in the automobile industry are quite different in the way that they are designed and how they operate and are usually referred to as secondary batteries or deep cell batteries. These batteries are usually available as lead acid batteries, which require scheduled maintenance, and captive electrolyte(gelcel) batteries which require no maintenance. The number of batteries required for a particular PV system will be a function of the desired capacity of the system such as 12 V, 24 V, 48 V or larger and this can be achieved by connecting a number of batteries in series and according to Lane (2021) there are currently four types of batteries available on the market that suitable for residential PV systems and these four include lead acid batteries, Lithium Ion batteries, Flow Batteries and Nickel based batteries inclusive of Sodium Nickel Chloride and Nickel Cadmium batteries. According to the International Renewable Energy Agency (IRENA) (2019) Brief, utility-scale batteries that provide power in the range of a few to hundreds of megawatt-hours (MWhs) are usually one of three types, Lithium-Ion Batteries, Sodium Sulfur Batteries and Lead Acid Batteries, with the Lithium-Ion Batteries currently dominating the world market for utility-scale batteries. Utility-scale Battery Energy Storage Systems (BESS) provide significant backup capacity to Variable Renewable Energy (VRE) systems such as solar PV and wind energy systems as they will be charged when energy is available and supply power to meet demands when energy from wind and sun is not available. The BESS system can also be used to provide clean renewable additional power to meet peak demand during peak periods when electrical power demands are high and thereby reduce or eliminate the need to operate additional dirty energy systems that negatively impact the environment.

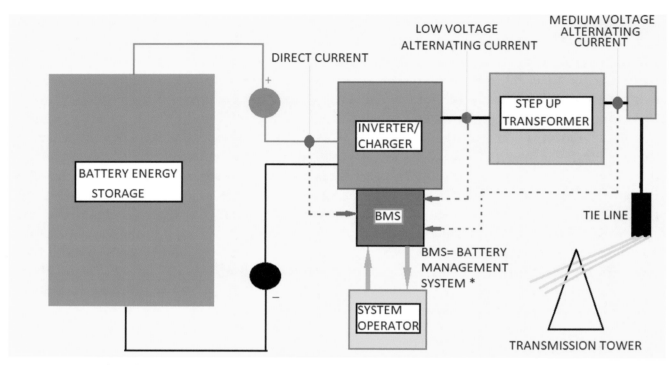

Fig 3.12 Typical Utility Scale Battery Energy Storage System. Source: NREL

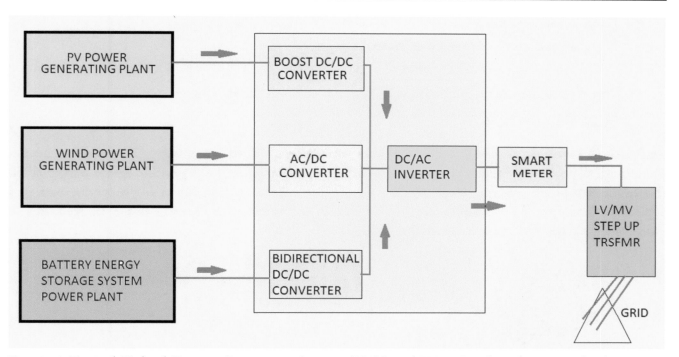

Fig. 3.13 Typical Hybrid Energy Generation System (BESS and VRE Combined Power Plant). Source: MISO Energy

Inverter. Photovoltaic systems generate electrical power in the form of direct currents(DC) only, similar to power produced by DC generators and batteries, while most electrical systems that serve residential, commercial, institutional and industrial facilities provide electrical power in a form known as alternating current (AC), based on the rotating equipment that produced this current and it therefore means that the DC power produced by PV systems is not suitable for direct use with most equipment and lighting systems used in residential, commercial, institutional and industrial facilities and is not compatible with the existing electrical supply standards and for it to be utilized in these and other such facilities the DC power would need to be transformed into the AC form of power and this transformation could only be done by using a device known as an Inverter or power conditioning unit(PCU). The basic types of inverters currently in use include Grid-Tie Utility Inverters, Stand-Alone inverters, and a Bi-Modal Inverter and according to Gomis-Bellmunt (2019) The basic function of the inverter are as follows:

- The conversion of DC power to AC power and voltage adaptation
- The tracking of the maximum power point

- Providing anti-islanding protection when connected to grid
- Providing synchronization when connected to a grid
- Grid disconnection, when connected to a grid
- Providing support to the grid when a PV system is connected to it.
- Integration and packaging

Stand-Alone Inverters. Stand-alone inverters work with either a PV array producing DC power or batteries producing DC power and would supply power to loads independent of the utility grid and based on their design they cannot synchronize with and feed power to the grid. They can convert 12V, 24V or 48V supply for residential applications and up to 480V for commercial and industrial applications and they can also be charged with an independent AC generator or other renewable sources.

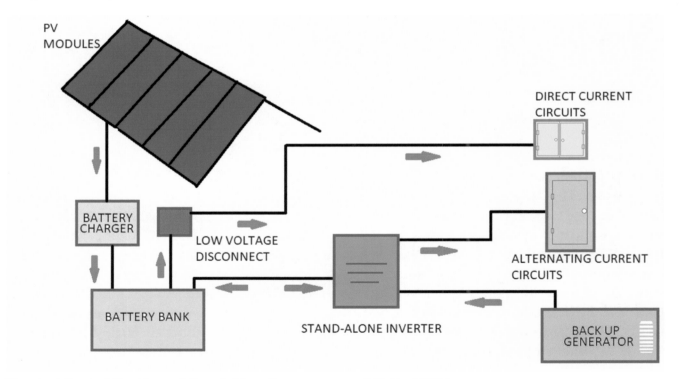

Fig. 3.14 Typical Residential Stand-Alone Inverter in an Off-Grid PV System.

Grid-Tie Interactive Inverter. These inverters operate from the PV array and supply utility-grade sine wave synchronized power to loads in parallel with the utility grid and they have the capacity to both supply internal loads and to export power to the utility grid. They are also designed with an anti-islanding safety feature which shuts down the PV system inclusive of the inverter when it senses a loss of grid power and will only restart after the grid has been reenergized. According to Dunlop (2012) there are several types of interactive inverters inclusive of:

- Module-Level Inverters. These are micro inverters that are usually attached to or are an integral part of PV modules with a rating of 200-300 watts.

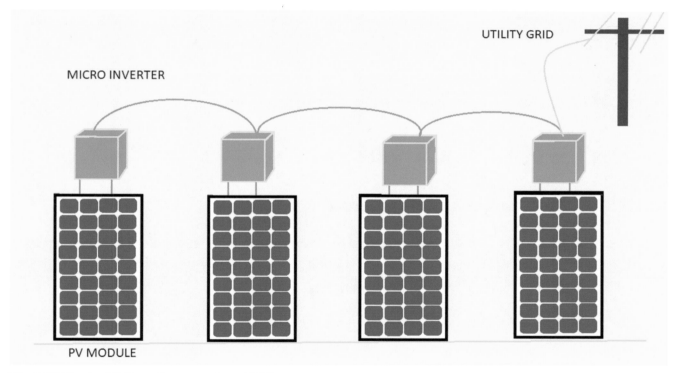

Fig 3.15 Typical Micro Inverters in a PV System.

- String-Level inverters. These inverters are designed for residential and small commercial PV systems which utilize 1-6 series connected module per string arrays and are usually rated at 2-12Kw.

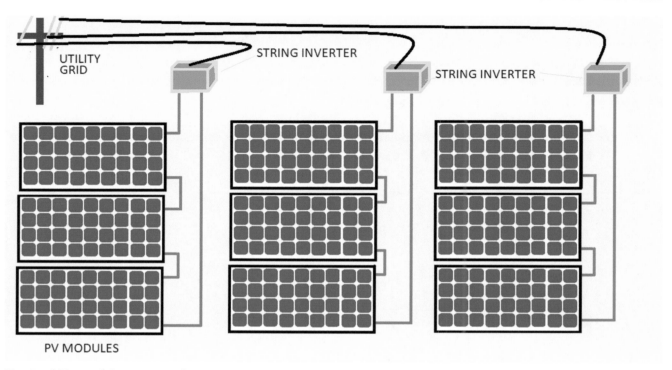

UTILITY GRID

STRING INVERTER

STRING INVERTER

PV MODULES

Fig 3.16 Typical String Level Inverters in a PV System.

- Central Inverters. These inverters are designed for commercial systems with multi-string arrays utilizing the same modules, same number of modules per string, the same circuit configuration, the same alignment and orientation and are usually rated at 30-500 Kw.

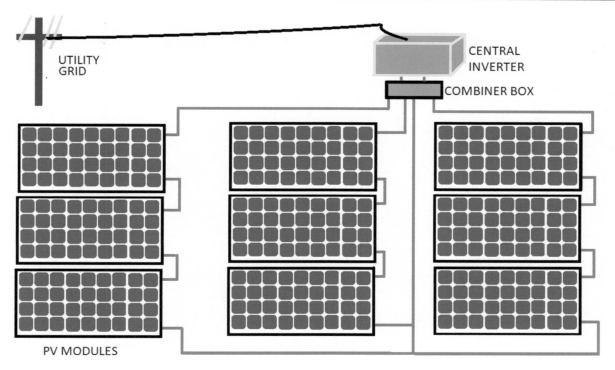

UTILITY GRID

CENTRAL INVERTER

COMBINER BOX

PV MODULES

Fig. 3.17 Typical Central Inverter PV System.

- Utility Scale Inverters. These inverters are designed specifically for solar farms in a utility scale PV power plant and are usually rated at 500KW - 1MW range and higher. These utility scale inverters include commercial grade inverters rated up to 500KW which are connected to the grid at voltages less than 600 VAC and utility grade inverters which are rated up to 1MW and higher which are used in large PV power plants which connect to the grid at distribution voltages up to 38 KV. These inverters normally use higher voltages, DC and AC, in order to reduce losses and the size of related cable and equipment, with the DC input voltages ranging from 900-1000 VDC and the output AC voltages ranging up to 38 KV after transformation. The larger utility scale inverters are usually presented as a packaged system which includes inverter, charge controller, switch gear, safety switches and climate monitoring and control system.

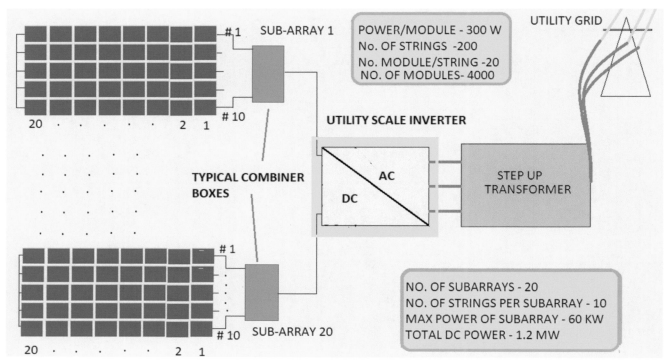

Fig 3.18 Typical Utility Scale Inverter in 1.2 MW PV Power Plant.

Bi-Modal or Battery-Based Interactive Inverter. This inverter charges batteries and produces AC power output that regulates the operation of the PV array when the grid is energized and transfers PV operations to a stand-alone mode when the grid is not energized to provide power to essential loads only.

In general inverters are designed for both single phase and multiphase operations and according to Dewangan and Nagdeve (2014) there are several classifications of inverters inclusive of:

- Single phase voltage source inverter
- Three phase bridge inverters
- Voltage Control in single phase inverter
- Pulse Width Modulated (PWM) Inverter

- Reduction of Harmonics Inverter, in inverter output voltage.
- Current source inverters

Modern inverters transform DC to AC by utilizing electronic switches and software that manipulate the direct current polarity so that it behaves in a manner similar to an alternating current and thereby making it suitable for adding to the grid and for typical household, commercial and industrial equipment. Typically, DC current is only suitable for computers, electronics and some lighting systems.

Maximum Power Point Tracking (MPPT)

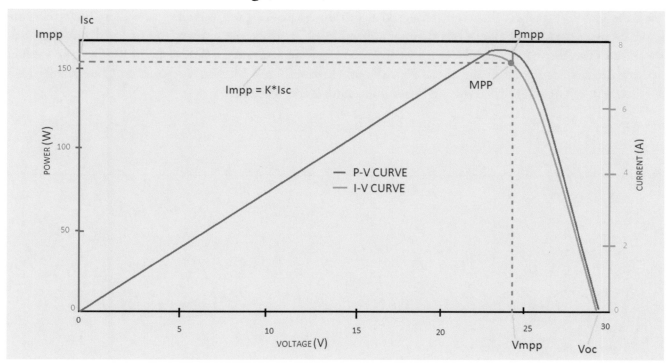

Fig 3.19 Typical Graphs Highlighting the Maximum Power Point of a PV System.

Standard operations of a PV system will generate current, voltage and power dependent upon the level of irradiance and the ambient temperature, with higher irradiance producing greater power output and higher temperatures reducing power output. The graph shown in Fig 3.19 highlights the relationship of current, voltage and power at a particular irradiance and temperature at a particular time of day and highlights the maximum power point (MPP) at this instance, however, the irradiance and temperature does not remain constant throughout the day and the maximum power point will fluctuate with the fluctuations on irradiance and temperature. This frequent uncontrolled fluctuation in power output from the PV array can be harmful to all downstream systems and in order to optimize operational output a control technique called the maximum power point tracking (MPPT) is applied to the system. The role of the MPP tracker is the continuous monitoring, measuring and adjusting necessary parameters that will allow the maximum power to be extracted from the PV plant irrespective of the weather, irradiance and temperature, or load conditions. The maximum power point of a PV plant changes continuously throughout the day and consequently the operating point of the PV plant will also change to optimize power produced and the MPPT will operate to ensure maximum output at all times. To continuously track the MPP the tracker utilizes one of several algorithms. According to Choudhary and Jain (2020) there are basically two categories of algorithms that are utilized in PV systems, the indirect and direct algorithms, with the indirect category consisting of Open Circuit PV Voltage Method and the Short Circuit PV Current Method and the direct category consisting of the Perturb and Observe(P&O) Method, Incremental Conductance (IC) Method, Fuzzy Logic (FL) Method and Neutral Network (NN) Method.

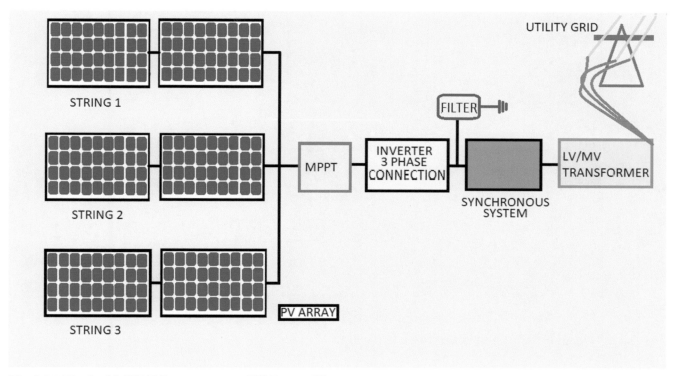

Fig 3.20 Typical MPPT Location in a PV Power Plant.

Perturb and Observe (P&O) Method. The P&O method algorithm measures the voltage and current of the PV panels, calculates the power and compares the resultant power with the previous calculated power and if the newly calculated power is greater than the previous one the controller maintains the same direction for the duty cycle, however, if it is less the controller changes the direction of the duty cycle, this process is relatively slow and time consuming. The P&O controller operates very well under steady operating conditions, however, frequent changes in environmental conditions, irradiance and temperature, will cause a significant reduction in its ability to produce at the optimum level. The performance of the P&O controller may however, be improved if it is used in combination with another controller such as the Fuzzy Logic Method controller (Choudhary and Jain 2020).

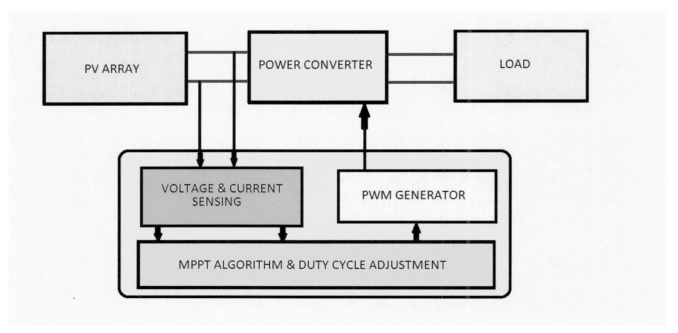

Fig. 3.21. Typical MPPT Controller Block Diagram.

Incremental Conductance (IC) Method. This IC method was developed to improve on the short coming of the P&O method as it reduces the tracking time and enhances power output when environmental conditions are unstable and are frequently changing. This algorithm takes into account and utilizes the current-voltage relationship to make necessary adjustments to the control element to achieve maximum power output.

Fuzzy Logic (FL) Methods. These are intelligent methods which gives an account of all relevant characteristics and qualities of the PV system using linguistic commands and the function of the components. These algorithms can model uncertainty and unpredictable situations in the PV system to formulate solutions to unstable power production and the FL method can also be combined with P&O method to greatly improve maximum power output from PV systems.

Artificial Neutral Network (ANN) Method. ANN methods are intelligent methods that have the ability and capacity to model PV systems utilizing available input and output data only and are known as black box systems which can model extreme nonlinear systems to produce more accurate estimates of PV systems maximum power output.

Location of MPPT. The MPPT may be a stand-alone device, be an integral part of the Charge Controller or be located within Inverters.

DC-DC Converter.

DC-DC converters are electronic devices which facilitate changing the levels of a DC voltage, either stepping up the voltage or stepping down the voltage as required, basically acting in a similar manner as alternating current, AC, transformers, basically making them DC transformers. DC-DC converters or DC transformer are vitally important as standard AC transformers do not facilitate level changes for direct current and voltage, a capacity that is of great importance to many pieces of electronic equipment. These converters are to be found in many pieces of equipment used in the PV power industry, computers and other electronic systems, EV and HEV car industry, trucks, other transportation systems and industrial systems.

The DC-DC Converter, while having no magnetic coils and metal parts that will cause hysteresis or heat transfer losses as occurs in AC transformers, are not 100% efficient as they utilized a portion of the incoming power to make the required level changes thereby ensuring that output power is always less than the input power. The efficiency in DC-DC converters range from 80-85% in the low efficient units to 90% in the more efficient units, indicating that anywhere from 10-20% of the incoming power is utilized to make the desired DC level changes. There are many types of DC-DC converters currently in general use inclusive of non-isolating converter, isolating converter, buck converter, boost converter, buck-boost converter, CUK converters, Change Pump converters, Flyback converters and forward converters. The focus however, shall be on the types which are usually used in PV systems, which includes the buck converter, boost converter and the buck-boost converter.

Buck Converter. Buck converters are designed to step down the supply voltage to a lower level while increasing current level as may be required in equipment such as charge controllers when the incoming voltage must be reduced to ensure effective charging of storage batteries. This reduction from input to output will usually give the converter a duty cycle of a fraction.

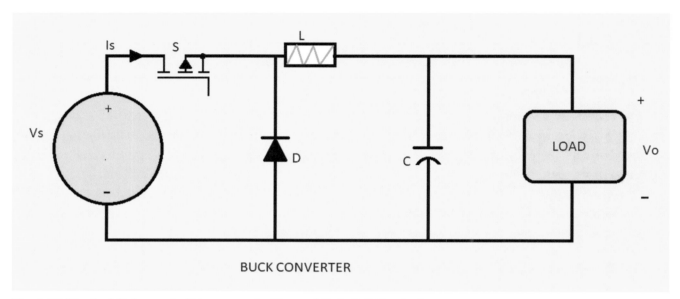

Fig 3.22 Typical Schematic Diagram of a Typical DC-DC Buck Converter.

The basic components of the buck converter as shown on the diagram are a switch, S, a diode, D, an inductor, L, a capacitor and a load/resistor, with V_s as the supply voltage and V_o the duty cycle being V_o/V_s, a fraction.

Boost Converter. The boost converters are designed to step up the supply voltage to a higher level as desired and giving the boost converter a duty cycle greater than one and these converters may be used in inverters which need to boost voltage to meet the voltage requirements of domestic household appliances.

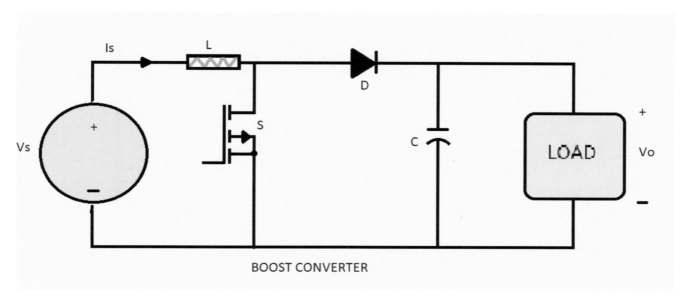

BOOST CONVERTER

Fig. 3.23 Typical Schematic Diagram of the DC-DC Boost Converter.

The basic components of the boost converter are inductor, L, capacitor, C, switch, S, diode, D, and load/resistor, the same as the buck converter but it has a different topography, as shown in the diagram. The duty cycle of the boost converter V_o/V_s will greater than one since the output voltage V_o has been stepped up to a higher level than the supply voltage V_s.

Buck-Boost Converter. Buck-Boost converters are a combination of the buck and boost converters and are designed to step down and step up the supply voltage depending on the duty cycle. In general, the buck-boost converter will step down the voltage when the duty cycle is less than 50% and step up the voltage when the duty cycle is greater than 50% and when the duty cycle is equal to 50% it will reverse the polarity of the output voltage.

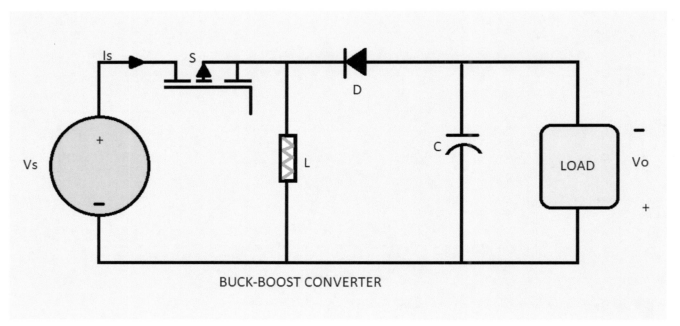

Fig 3.24 Typical Schematic Diagram of the DC-DC Buck-Boost

The basic components of the buck-boost converter are inductor, L, capacitor, C, switch, S, and load/resistor, similar to both the buck and boost converters but with a very different topography, as shown in the diagram. The duty cycle of this converter varies significantly dependent on load conditions and may be used in MPPT systems. In general DC-DC converters are widely used in all PV systems electronics inclusive of charge controllers, inverters, MPPT, power optimizers and are an integral part of the MISO Power Platform.

PV System General Wiring

The construction of all PV system larger than a single cell is integrally dependent upon systems of wiring and interconnections, with the methods of wiring and interconnection playing a significant role in the voltage and current output for each system. The basic wiring of PV systems commences with the terminals on each cell and extends outwards to the interconnecting of cells to make modules, and the interconnection of modules to make arrays. The module/panels connections are followed by connections to PCU, charger controllers, inverters, disconnects, batteries and the final connections to grids and or loads. According to Sheehan and Coddington

(NREL)(2014) there are three types of PV System connections with each system requiring different wiring requirements as specified by the NEC and the 3 types are:

- Grid -Tie
- Grid-Tie with battery backup
- Stand Alone

Grid-Tie. The grid-tie PV system is the simplest and lowest cost PV power supply system and it has the capacity to displace utility supply, supply power to and receive power from the utility through net-metering and it does not utilize battery storage or any other form back-up power. The components of the grid-tie system are, PV for converting sunlight into electricity, grid-tie Inverter for converting DC from array to AC for local use and sending to utility grid and a safety switch which allows the utility to shut down the system in an emergency.

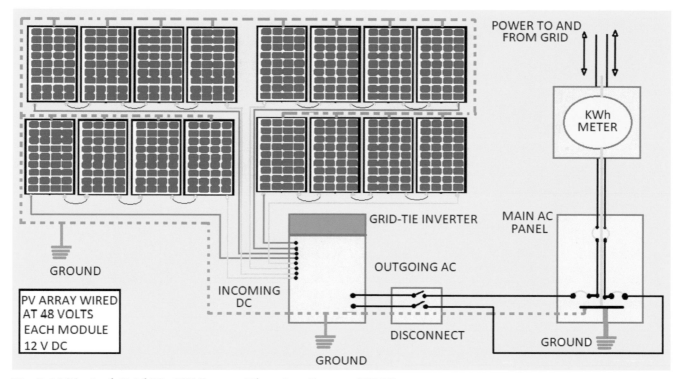

Fig 3.25 Typical Grid-Tie PV System Electrics. Source: NREL

Grid-Tie with Battery Back Up. This arrangement allows for greater flexibility in managing available power as the back-up battery provides a limited supply for critical load in the event of a power outage. The one drawback to a battery back-up system is usually the need to maintain batteries dependent on the type of battery used, specifically lead acid batteries. The main components of this system are similar to the Grid Tie System with the exception of the battery for storage of electricity for night time or emergency use and the battery cutoff switch which is used to isolate the battery for maintenance or in the case of an emergency.

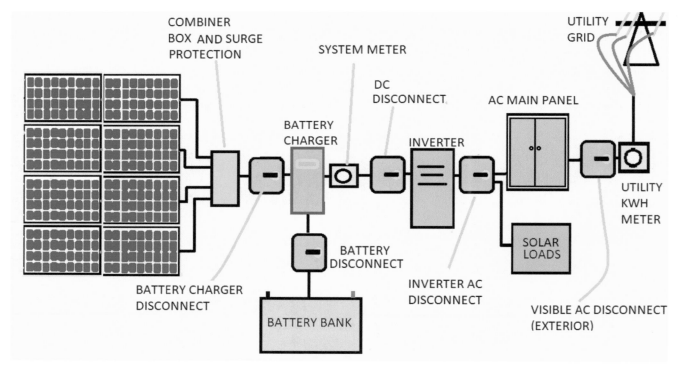

Fig. 3. 26 Typical Grid-Tie PV System with Battery Back-Up.

Commercial Stand-Alone PV Systems

Stand-alone PV systems are usually utilized to provide electrical power to isolated places and equipment inclusive of homes, commercial businesses and institutional facilities, these systems however do not supply power to the grid but may utilize the grid as back-up source of power. Stand-alone PV systems are usually utilized for lighting, indoor and outdoor, and small appliances that typically utilized in all types of residential facilities and the main components of such systems are solar array, batteries, battery charge controller, a stand-alone AC inverter, and a back-up power supply.

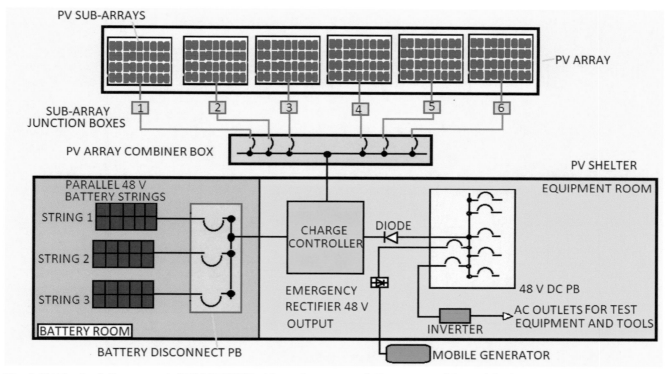

Fig 3.27 Typical Commercial PV STAND Alone System with Battery and Portable Generator. Source: NREL

Wiring of Battery Bank in PV System

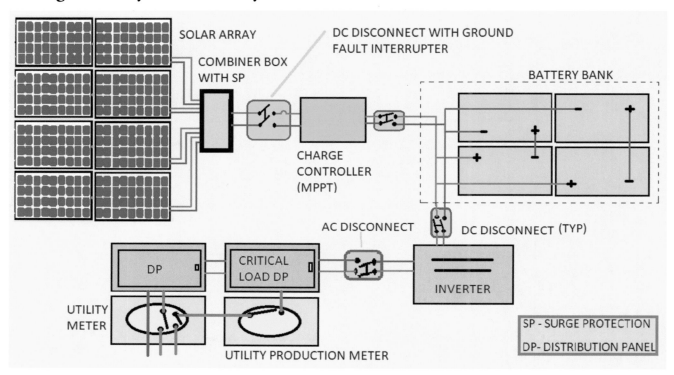

Fig. 3.28 Typical PV and Battery Wiring Diagram. Source: NREL

Photovoltaic Module Wiring

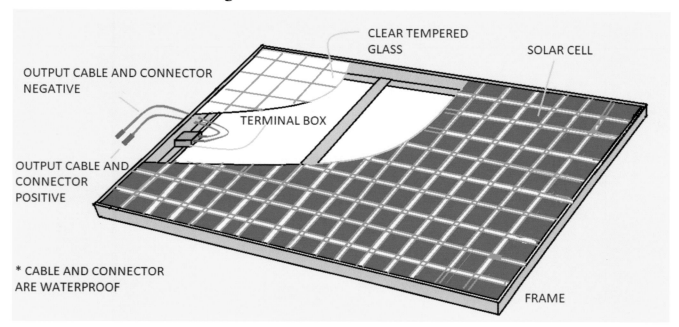

Fig. 3.29 Typical PV Module Electrical Terminal.

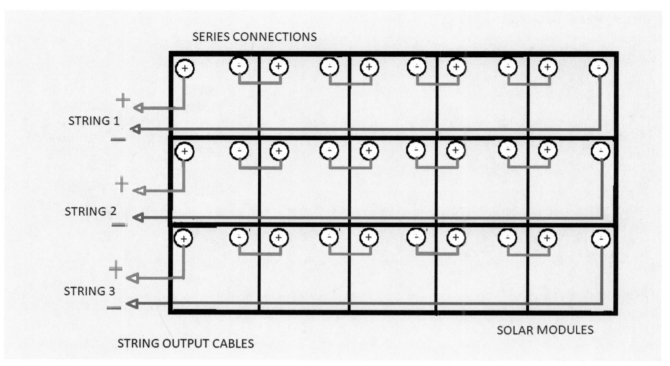

Fig. 3.30 Typical PV Array Series Wiring.

PV modules are typically wired in series, parallel or a combination of series and parallel as determined by the desired voltage and current output with the installations in series increasing voltage while the installations in parallel increases current and modules are typically arranged as shown in the diagrams below.

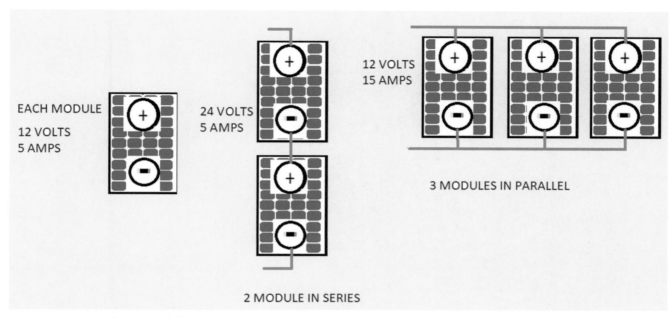

EACH MODULE

12 VOLTS
5 AMPS

24 VOLTS
5 AMPS

12 VOLTS
15 AMPS

3 MODULES IN PARALLEL

2 MODULE IN SERIES

Fig. 3.31 Typical PV Modules Wired in Series and Parallel.

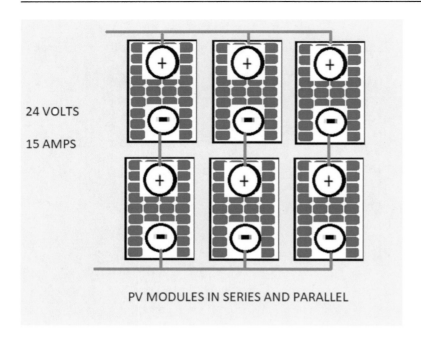

24 VOLTS

15 AMPS

PV MODULES IN SERIES AND PARALLEL

Fig. 3.32 Typical PV Modules Combination Series-Parallel Wiring.

The wiring of all PV array electrical systems is dependent on the number of modules in series or parallel, the desired voltage and current, the maximum voltage and current and the system short circuit current. The wiring system, wires and connections, used to connect the modules are usually PV specific wires that are specifically designed to withstand temperatures between 90^0 C in wet conditions to 150^0C in dry conditions. These wires are usually made from stranded copper or aluminum wire to give desired flexibility, are covered with thicker than standard insulation to provide additional protection from the sunlight and must past two tests, flame testing and strict sunlight resistance testing.

Combiner Boxes. Combiner boxes are typically installed directly between the modules and the inverter making them the first piece of electrical equipment to interface with the output of the modules and according to Smalley (2015) the purpose of the combiner box is to gather the output wiring of several strings of modules together. The output wiring of each module string is connected to a fuse terminal and the output from each fused terminal are brought together into a single larger conductor which connects the combiner box with the inverter. These combiner boxes also incorporate overcurrent and overvoltage protection switches, disconnect switches and monitoring equipment. Typically, while combiner boxes may provide some benefits for smaller 2-3 string of modules in residential installations, they are not a requirement, however, combiner boxes are required for larger commercial and utility scale installation of 4- 4000 strings.

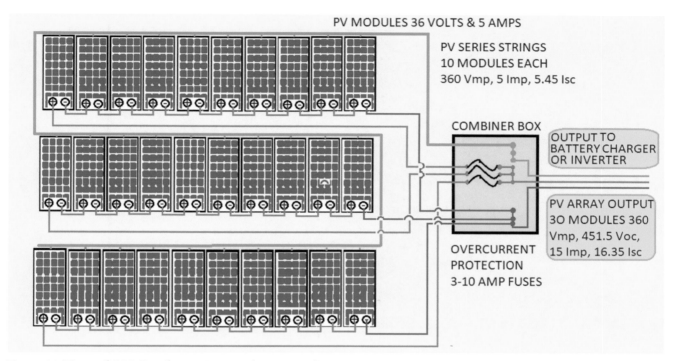

Fig.3.33 Typical PV Combiner Box and PV Panels Arrangement.

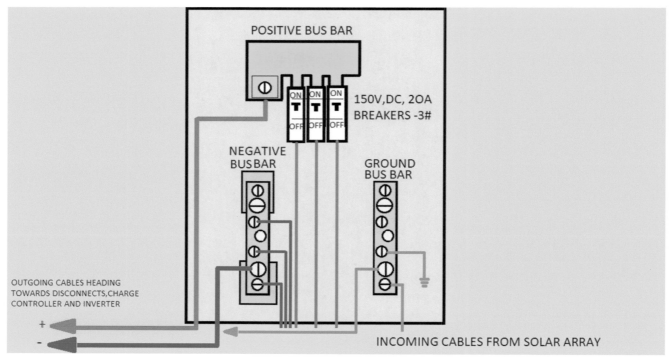

POSITIVE BUS BAR

ON ON ON
OFF OFF OFF

150V,DC, 2OA
BREAKERS -3#

NEGATIVE
BUS BAR

GROUND
BUS BAR

OUTGOING CABLES HEADING
TOWARDS DISCONNECTS,CHARGE
CONTROLLER AND INVERTER

+

-

INCOMING CABLES FROM SOLAR ARRAY

Fig. 3.34 Typical Internals Of A Combiner Box.

Power Optimizer

The purpose of the power optimizer is to reduce the output voltage and increase the output current to allow for more panels to be connected in series to match the capacity of the attached centralized string inverter. The basic functions of the Power Optimizer are as follows:

- Monitoring of each panel's performance (MPPT)

- Continuous comparison of the performance of each panel in a string

- To optimize the performance of each panel with different orientations relative to each other by preventing panels from being impacted by problems associated with the other panels, thereby reducing greatly the problems associated with shading and allowing for a greater density, up to 20%, of panels in an array on a particular area of land and thereby increasing overall power output of a PV power plant.

- "Conditioning" the output from each panel by optimizing the voltage before it reaches the string inverters downstream.

To be effective one power optimizer must be installed on each panel and must be paired with a centralized inverter, with more modern state of the art optimization occurring at the cell-string level which ensures an increase in the output from each panel significantly. The state-of-the art technology and cell-string level optimization facilitates a significant improvement in performance and a significant reduction in material and labor cost as components such as the optimizer box, additional wiring, communications devices, internet connections and proprietary inverters are eliminated.

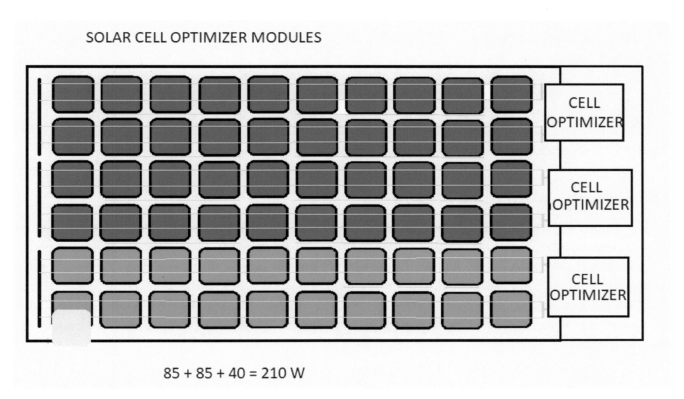

SOLAR CELL OPTIMIZER MODULES

85 + 85 + 40 = 210 W

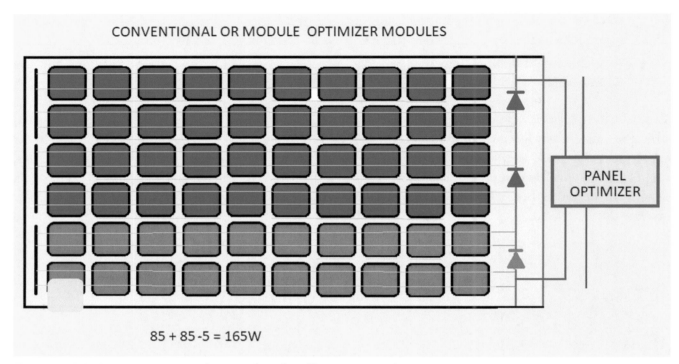

CONVENTIONAL OR MODULE OPTIMIZER MODULES

PANEL OPTIMIZER

85 + 85 -5 = 165W

Fig. 3.35 PV Optimization at the Panel and Cell-String Levels.

PV Mounting and Tracking Systems

All PV systems require the use of support systems that will hold the PV modules in a fixed position or allow them to continuously follow the movement of the sun from sunrise until sunset to optimize the amount of energy collected and converted on a daily basis and these support systems may be fixed mounting racks which point the modules in a southerly direction all the time, mounting poles which hold the module is a fixed southerly direction, poles or racks with single axis tracking systems in which all the modules the track the sun in an east to west direction in its travel across the sky and poles or racks with a dual axis tracking system which allows the modules to track the movements of the sun east to west and north to south.

Fixed Mounting Racks. Fixed mounting racks can be located on roofs or on the ground with the amount and type of fabrication and construction work required varying to meet the particular need of the location. The roof mounted systems are usually fabricated using special designed roof attachment systems and necessary waterproofing to ensure that the integrity of the roof is not harmed in any way. The ground mounted systems are fabricated using shaped hollow steel sections and high strength reinforce concrete bases and foundations that will keep the attached panels stable during all weather-related events. For optimum energy production the panels, on roofs and on the ground, must all be set at an angle, tilt angle, relative to the sun to ensure perpendicular impingement of the rays of the sun on the solar panels at high noon and this angle is usually location dependent based on the latitude of the location in which the panels are to be installed.

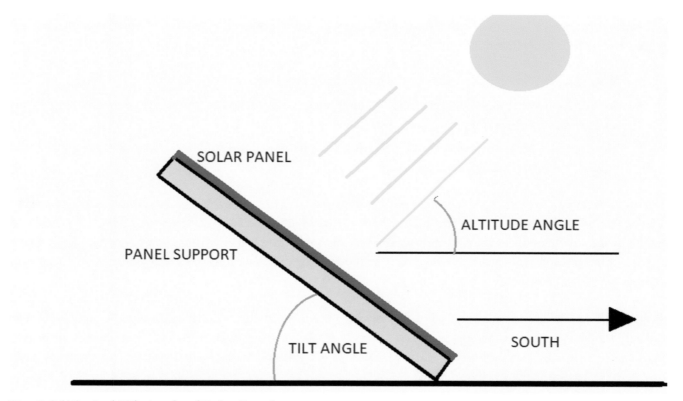

Fig. 3.36 Typical Tilt Angle of Solar Panels.

The fixed frames will therefore be fabricated to ensure that the bed on which the panels will be installed will be set at the tilt angle relative to the base of the frame located on the ground and on the roof.

PV Systems Solar Tracking Systems. Solar tracking system can generally be divided into two categories passive and the active tracking systems, with the passive tracking system utilizing the heat from the sun and a gas contained in an enclosure with attached linkages to move the panels as the gas expands and contracts when heated and cooled throughout the day. The active systems on the other hand require the input of an external source of electrical energy, DC, to drive electrical motors and associated mechanical systems to move attached PV panels in the required directions. In general, most PV solar systems utilize active tracking systems and there are two major active tracking systems that are in general use, single-axis tracking and dual axis tracking systems and both of these tracking systems significantly improves the energy output of attached PV systems. According to Marsh (2021) the single-axis tracker improves output from PV panels by as much as 25-35% while the dual-axis tracker improves the output from PV panels by as much 45%. The general benefits of utilizing solar tracking systems are as follows (Marsh 2021)

- Solar systems that track the path of the sun generate more electricity
- The production of a significantly greater amount of electricity per panel will reduce the number of panels required for a particular application.
- The improved production of electricity by a grid-tie system with a variable electricity rate agreement could be more advantageous economically.

The main drawback of utilizing trackers on a PV system are the increased capital costs and the increased cost to operate and maintain the systems as the trackers have a relatively high first cost and they utilize both electrical and mechanical parts that will be affected by wear and tear caused by external conditions and the regular movements of parts. One other disadvantage of using trackers is that they are fairly large and heavy and are not normally suitable for use on most roofs.

Single-Axis Tracking System. The single axis tracking systems normally have their axis aligned north and south to facilitate the east to west movement of the panels to track the movement of the sun from sunrise to sunset. Single axis systems are used extensively on large power plant applications that have large land space to work with.

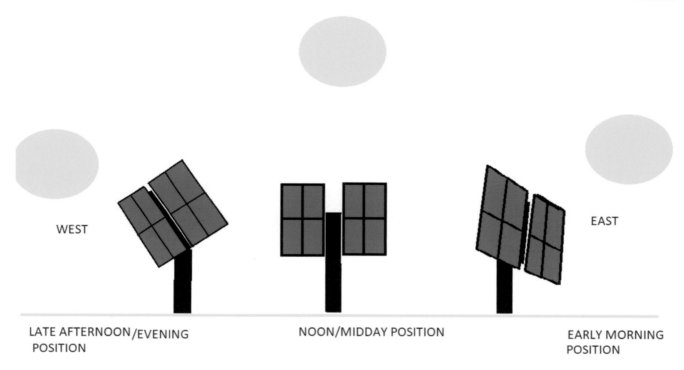

WEST

EAST

LATE AFTERNOON/EVENING
POSITION

NOON/MIDDAY POSITION

EARLY MORNING
POSITION

Fig. 3.37 Typical PV System East to West Tracking of the Sun Using Single-Axis Tracker.

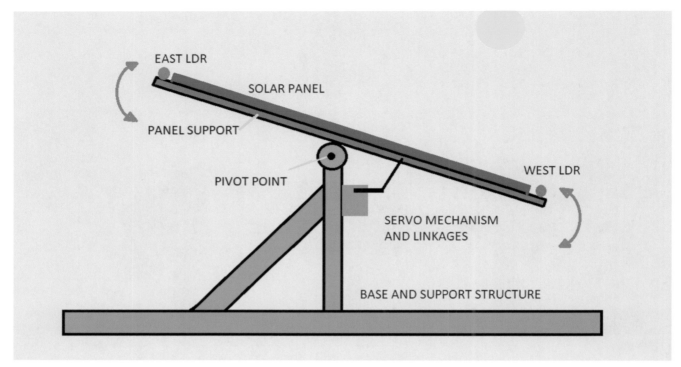

Fig. 3.38 Typical Schematic Highlighting Single-Axis Tracker Components.

Dual Axis Tracking System. The dual-axis tracking system has it axes aligned both north and south and east and west to facilitate the normal east to west movements and the seasonal variations in the height of the sun. The dual-axis tracking systems find great applications on projects with limited space and with concentrated photovoltaic systems.

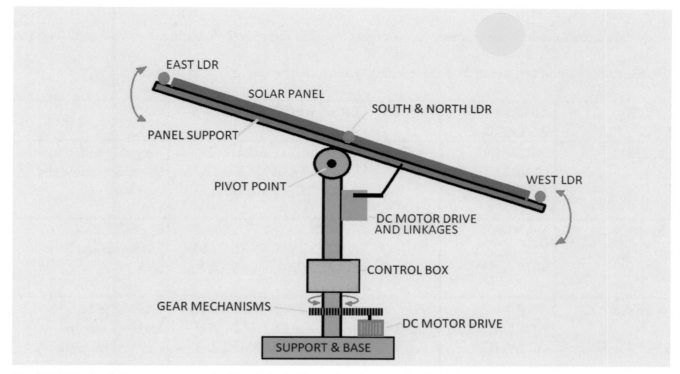

Fig.3.39 Typical Components of a Dual-Axis Tracking System for PV Systems.

Components of Tracking Systems

The basic components of a tracking system, single and dual axis, includes the following:

- Arduino Uno Microcontroller (or similar microcontroller)
- DC motor (2- for Dual-Axis) and DC motor driving cards
- Relays and relay cards
- Solar radiation Sensors
- Solar position sensors
- Wind Sensor and anemometer
- Light Sensors (LDR)

- Ice and hail detectors
- Associated mechanical systems inclusive of gear box, gear and pinion, shafts, linear actuators and linkages.

The basic benefits and requirements of the tracking systems are as shown below in the table.

TYPE	AXIS OF ROTATION	YIELD	AREA REQUIRED(MWp)	O&M
FIXED	FIXED IN HORIZONTAL AND VERTICAL	BASE	4-5 ACRES (DEPENDING ON MODULE LOCATION AND TILT ANGLE)	Very easy with the lowest downtime and basic cost
SEASONAL	MANUAL VERTICAL ROTATION	8-9 % HIGH	4-5 ACRES (DEPENDING ON LOCATION IT CAN HAVE SIMILAR AREA AS FIXED)	Easy with lowest downtime and basic cost
SINGLE AXIS	AUTO HORIZONTAL ROTATION	25% HIGH	5-7 ACRES (DEPENDING ON LOCATION IT CAN HAVE SIMILAR AREA AS FIXED)	Difficult with higher downtime and 10-15% higher basic costs
DUAL AXIS	AUTO HORIZONTAL AND VERTICAL ROTATION	30% HIGH	7-9 ACRES	Difficult with higher downtime and 20% higher basic costs

Table 3.4 Typical Tracking Systems Data.

PV Systems Grid Connections

PV systems produce electricity that may be used in stand-alone systems, grid-tie systems, grid-tie with battery backup systems, grid-tie with auxiliary power back or used as a combination of systems for residential, commercial, institutional, industrial and utility purposes. The production of the electricity for the grid is usually done using many electrical power generating technologies inclusive of the combustion of fossil fuels, the utilization of water to create hydropower, the utilization of nuclear energy, the utilization of the natural heat in the earth to create geothermal energy, the utilization of energy contained in wind to create wind energy systems, onshore and offshore, the utilization of radiant solar energy in five concentrated thermal energy systems and the utilization of the photovoltaic effect created by photons of light impinging on certain material to create two Photovoltaics (PV) energy systems, PV and CPV. As with all other energy systems the electrical power produced from PV systems must be interconnected to an existing or new power distribution systems and the connection to the grid is usually facilitated by interactive inverters which are designed to communicate with the grid, of which there are several types as discussed earlier. Three of the most common grid connections are the residential, commercial and utility scale power plant, all of which utilize specific inverters to facilitate connection to the grid.

Single Family PV System Grid Connections

There are three possible arrangements for a grid connected single family residential PV system is as shown in Fig 3.40 below with the interactive inverter supplying power to the residence through a panel box and to the grid through a smart meter which allows the flow of electrical power in two directions and net metering which will allow the home owner to supply excess power to electrical power distribution system and to allow them to be paid for this excess power based on power supply agreement with the owners of the distribution system.

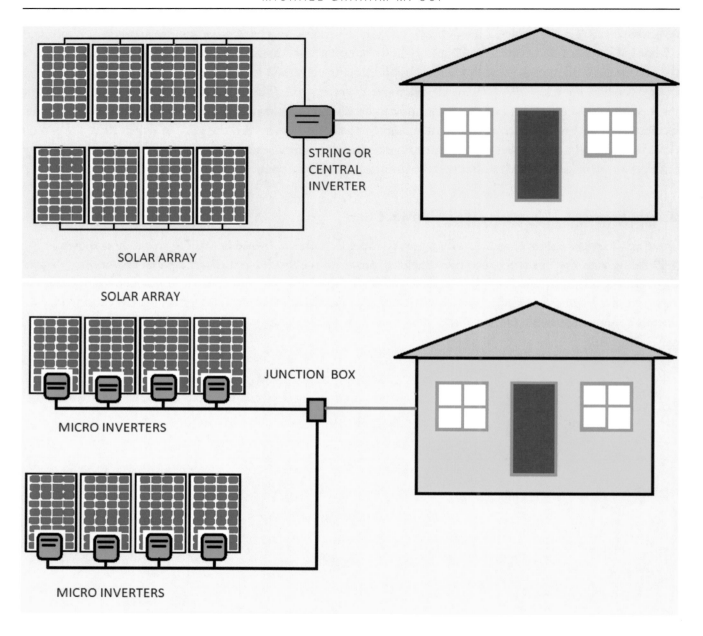

SOLAR ARRAY

STRING OR
CENTRAL
INVERTER

SOLAR ARRAY

JUNCTION BOX

MICRO INVERTERS

MICRO INVERTERS

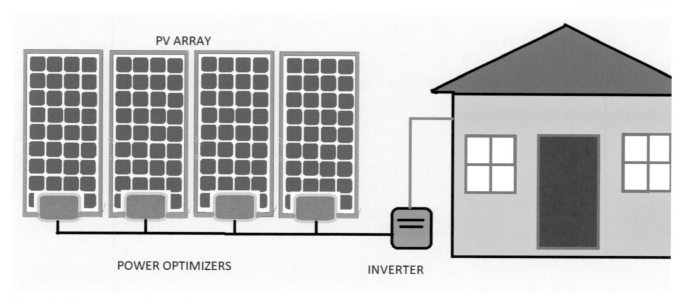

Fig 3.40 Typical Residential Inverter Arrangements.

According to Texas Instrument the interactive inverter utilized in grid connected single family residences must have certain characteristics inclusive of:

- Residential Interactive Inverters are usually string inverters
- Must be single phase 120/240 Volts system
- Must have the capacity to convert direct current (DC) to sinusoidal alternating current (AC). Must create an almost true sine wave to match grid system.
- Must have the capacity to track the Maximum Power Point.
- Must have the capacity to maintain a constant output voltage and frequency on the systems within a given voltage and frequency range and provide voltage and frequency synchronization with the grid.
- Must be large enough to supply power required by all household equipment at the same time and have the ability to control surges cause by the use of equipment like dishwashing machines, clothes washing and drying machines that are used intermittently.
- Must have capacity to send excess power to the low voltage side of the grid.
- Must have the capacity to continuously monitor grid voltage and frequency levels according to IEEE 1547.

- Must have the capacity to disconnect from grid when grid voltage and frequency levels fluctuate outside of predetermined voltage and frequency levels/ranges for a predetermined period of time (IEEE2003 & IEEE 2014) or when grid power is lost, islanding protection.

- Must have the capacity to "ride through" small fluctuations in voltage and frequency

- Must have the capacity to control the real and reactive power output of the distributed system by changing the level of real power output from the system by controlling the rate at which real power is fed into the grid.

- Must have the capacity to prevent the reoccurrence of grid disturbances by delaying the return of the PV system to the grid for a fixed period of time after grid power has returned after an outage.

- Must have elaborate monitoring and communication with the grid and have the ability to receive instructions from a central location.

- Must have the capacity to make autonomous decisions to improve grid quality, support power quality, control harmonics, and provide required additional ancillary services.

- Must provide galvanic isolation by preventing the injection of DC power into the grid.

Commercial PV System Grid Connection

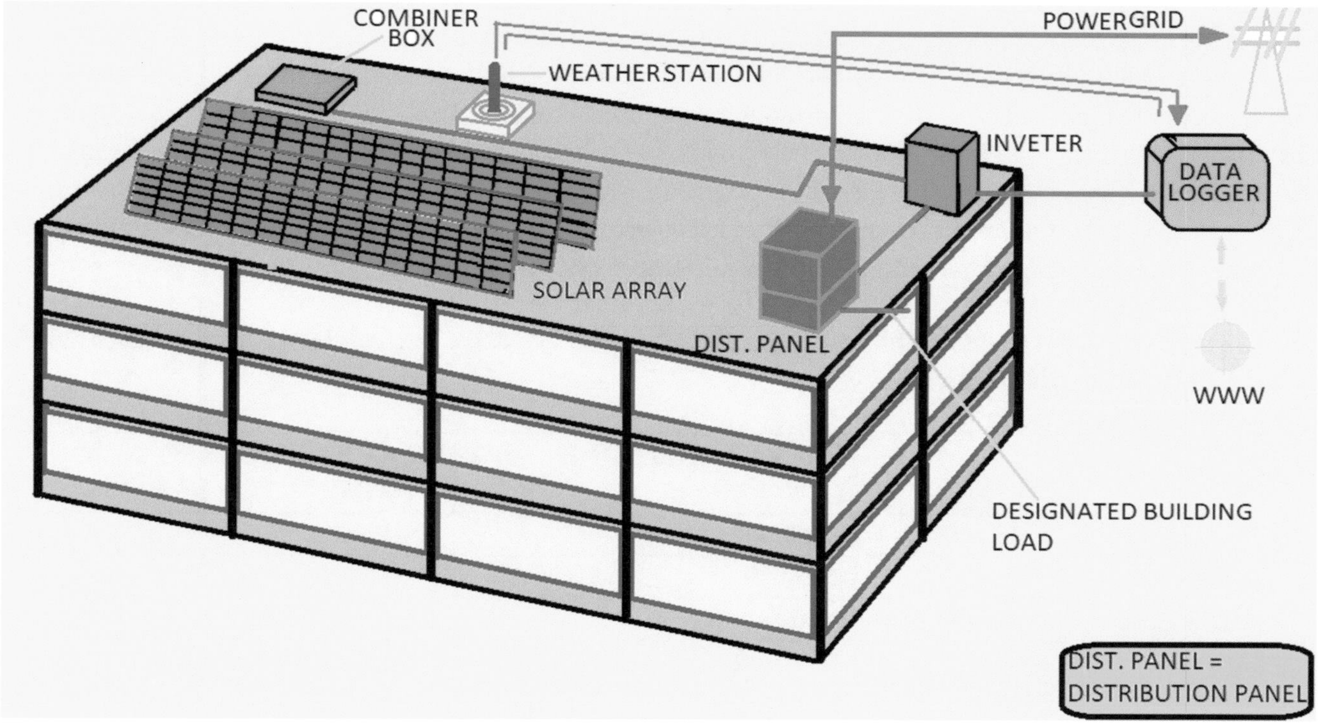

Fig 3.41 Typical Centralized Inverter for Commercial Grid-Connected PV System.

Commercial solar systems typically use 3- phase 277/480 volts inverters arranged in one of two configurations, centralized or decentralized, with the decentralized arrangement utilizing an inverter manager which acts as an interface between all the inverters. The inverters used in commercial systems can be installed on the roof, for roof mounted systems or on the ground for systems mounted on the ground.

Centralized Versus Decentralized Inverters

The use of centralized inverter versus the use of decentralized inverters is usually based upon several factors inclusive of, the initial cost to install, the reliability of each system and the maintenance cost. In general, the standard rule of thumb that has been used is that large, commercial, and very large, Utility Scale, PV power plants function much better when they utilize large centralized inverters and that smaller PV plants, residential and small commercial, function much better with decentralized inverters, however recent evidence is indicating that decentralized inverters can work just as well as in large and very large PV power plants without any significant disadvantage. According to a report prepared by Tanomvorasin and Thanarak (2015) a comparison of two similar sized power plants,1MW and 1.2MW, with one using centralized inverters and the other using decentralized inverters, in Thailand, indicates that the reliability of the decentralized plant (99.98%) is usually a little higher than that of the centralized plant (97.47%) based on the length of time taken to repair or replace faulty inverters in both plants. The replacement time is a function of the availability of small, lightweight inverters that can be easily replaced by an operator or technician in an hour versus the longer time required 5-8 hours to get experts to work on the much larger more complexed central inverter system.

The second element of concern is the cost to install the different systems, the major cost drivers of each system is highlighted in the table below.

ITEM	DECENTRALIZED INVERTER US $	CENTRALIZED INVERTER US $
AC WIRES	5,101.00	0.00
DC WIRES	7,569.00	23,846.00
COMBINER BOX	0.00	10,000.00
AC BREAKER	11,417.00	5,500.00
DC BREAKER /PV FUSE	1,440.00	10,600.00

Table3.5 The Comparative Cost of The Major Cost Drivers in A 1MW PV Plant Using Centralized and Decentralized Inverter Installations. Source: wernermn.com

The information given in the Table 3.5 above is general information for 1 MW PV power plants and is not related to the power plants in Thailand, referred to above. The information shown indicates that in general, the costs for direct current (DC) related wires, combiner boxes and breakers make the installation of a centralized inverter system much more expensive than for a decentralized inverter system for a similar sized plant.

The third important element, maintenance, is of great importance over the projected life of 25-30 years for a PV plant, as a plant with a larger number of equipment suggest the requirement for far more effort and money to keep all systems functioning at required optimum levels. The number of inverters on the two power plants in Thailand were, 255 on the decentralized plant and 2 on the centralized plant, and these numbers may suggest that while it may be easier to replace a single small inverter (1 hour) versus repairing a large inverter (5-8 hours), the maintenance program for the decentralized plant with 255 inverters should be much more complexed than that for the centralized plant with only 2 inverters with the level of complexity being related to the logistic required to ensure the ready availability of replacement units that will allow the quick replacement of a small broken inverter in one hour. The immediate availability of the replacement inverters would require that the facility carry a fairly high inventory of these inverters, with associated inventory costs. This inventory and related carrying cost and the total time to replace more than one small inverter during a year could make the repair of one more robust large inverter much more cost effective. Similarly, the cost to troubleshoot 255 inverters with related accessories and circuits versus troubleshooting 2 inverters with related accessories and circuits could be much greater as even with remote troubleshooting, no wiring, there would be a need for a more complexed software and hardware system to manage 255 units versus 2 units.

Components of a Commercial Inverter

Most commercial inverters are fairly complex and typically consist of three sections, inverter- DC-AC transformation, power optimizers, monitoring platform, depending on the manufacturer.

Inverter Section. This section of the commercial inverter is designed for three phase 277/480 Volts power, operating at high efficiencies, typically 98.8%, with capacities ranging from 15-100 KVA. They are designed to work with power optimizers, contain a Maximum Power Point Tracker (MPPT) and a wireless communication system with the option to plug in smart phones. They are also provided with built-in electrical protection systems inclusive of direct current (DC) safety switches, surge protection for both alternating current (AC) and direct current (DC) power trains. In some designs the utilization of a decentralized inverter system may require the use of an inverter manager, which is a central communications component with the role of acting as the only interface between all the inverters and for controlling all important inverter and system management functions.

Power Optimizers. These optimizers perform the following functions:

- Module level Maximum Power Point Tracking (MPPT) to eliminate mismatch power losses
- Facilitation of the use of module strings of uneven lengths and modules installed at varying azimuth and tilt angles.
- Provision of an automatic protection system for each module that will disconnect each module as required.

Monitoring Platform. The Monitoring Platform consist of three major elements each of which perform essential function for the inverter and these elements are commercial gateway, performance monitoring and grid interaction. The overall role of this platform is to provide:

- Provide an overview of all system performance
- To troubleshoot all system components remotely
- To provide direct access to all systems through the internet, by using any browser, laptop computer, smart phone or tablet.
- To provide continuous communications with all power optimizers through the DC power lines.

Commercial Gateway. This is an electronic data collecting device that is connected to the weather station which has multiple environmental sensors inclusive of the irradiance sensor, solar position sensor, temperature sensor, wind sensor, hail sensor, which are all used to collect data that is utilized in analyzing system performance.

Performance Monitoring. This a central processing unit, CPU, that is used to calculate performance ratios for specific locations and to assess and evaluate environmental conditions based on the data collected from the sensors or from a satellite-based service.

Grid Interaction. These are monitoring and control systems that supports power control inclusive of grid synchronization, real and remote reactive power, secondary grid protection by inverter relay control, AC, and low voltage and frequency ride through.

Utility Scale PV Inverter Grid Connection

Inverters used in utility scale solar plants can either be decentralized or centralized inverter depending on the scale of the project. For the smaller plants a case may be made for using the smaller decentralized inverters while at the higher end the "Mega-Watt in a Box" central inverters are the preferred choice.

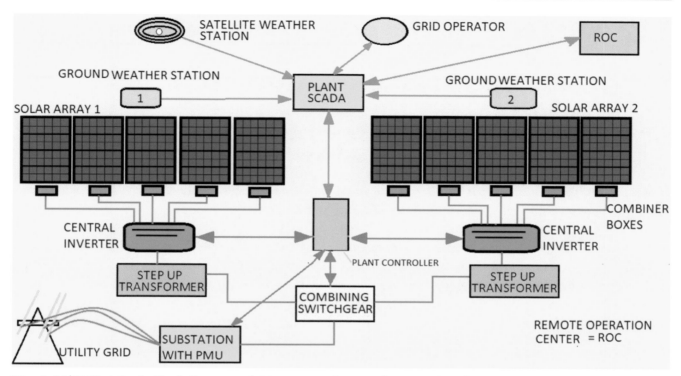

Fig. 3.42 Utility Scale Grid Connected PV Power Plant with Centralized Inverters.

Utility scale PV power plants sizes range from 5 MW to hundreds of MWs and the inverter systems utilized by these plants are usually a function of efficiency that can be achieved, total output required and the overall cost to install and maintain them over the life of the inverter equipment. As earlier indicated large commercial and utility scale power plants can use either a centralized or decentralized inverter system without any significant advantage or disadvantage. However, on the much larger plants there are several factors that can greatly influence the choice of inverter selected and these factors must be given due consideration in the selection process and these factors include Reactive Power Control, Real Power Curtailment and Power Factor Control. Apart from these factors, there are also five key metrics that are of vital importance to the efficient and effective operation of the utility scale power plant over the life of its combined systems and each metric must be monitored and controlled, these metrics include basic performance, capex (Labor and Material), system operating efficiency, inverter service life and true cost of service.

Basic Performance.

Basic performance covers:

- The ability to connect systems at distribution voltage, 12.47 – 34.5 KV, levels and transmission voltage, 42- 230 KV, levels.
- Grid support functions inclusive of Voltage Regulation and VAR support
- Low Voltage Ride Through (LVRT) and High Voltage Ride Through (HVRT)

Connectivity. The connectivity to the distribution and transmission systems voltages are usually provided by the step-up transformers and substations located within close proximity to the power plant.

Grid support. The grid support functions require closed loop control through the power plant control (PPC) system and from a communications architecture point of view central inverters are much easier to coordinate with the PPC since it has fewer inverters to communicate with and also has the capacity to utilize fiber optics instead of copper wire between inverters. The central inverter also has the ability to deliver superior Reactive Power (VAR) Control support relative to string inverters which are typically limited to plus or minus 0.8 power factor while the central inverters offer a much wider range of power factors plus or minus 0.8-0.9

Voltage Ride Through. Central Inverters provide superior voltage ride through capabilities as it allows 1.3- 1.4 PU for HVRT faults while string inverters have a limit of 1.1- .12 PU. The central inverters are also provided with complex ride through curves that can be programmed while the string inverters are limited to one or two curves which makes it more difficult to coordinate with complex grid interconnection requirements.

Capex

According to Snyder (2011) string inverters require 2.2 cents/W more than central inverters in a utility scale system, 1.9 cents/W more for material and 0.3 cents/W more for labor.

System Operating Efficiency

Inverter systems efficiencies are as follows:

- Central and String Inverters have the same electrical losses, 1% in the DC systems and 0.15% or negligible in the AC systems

- Greater electrical losses, 0.85%, in utility scale power plants designed with string inverters than with plants designed with central inverters.

Inverter Service Life

- The manufacturers of central inverters for utility scale power plants usually offer central inverters with the same service life as the power plant 25-30 years.

- The central inverters were usually provided with a proactive service and maintenance program that is designed to ensure the designed longevity, 25-30 years. To ensure the success of these service and maintenance programs the supplier would provide service contract which is based on limiting plant downtime and thereby guaranteeing a specific plant up time. These service contracts significantly reduce the risks associated with service and maintenance and thereby control losses due to downtime.

In contrast to the stability offered by central inverters string inverters utilized in a decentralized system have the following disadvantages:

- No published or established service life and maintenance usually involves the replacement of major components or the complete replacement of a unit

- Replacement of major components or the replacement of a complete string inverter may not be feasible 10-15 years after installation due to the many technological changes which would have taken place during that period of time, changes such as the fact that old technology string inverters utilized 600V while newer string inverter may utilize 1000-1500V.

In general, central inverter have service contracts with guaranteed uptimes that are available from the most reliable suppliers, while the risks and future service cost for string inverters can be said to be indeterminate.

True Cost of Service

The true cost for any piece of equipment is always a function of the capital cost and associated maintenance cost over the projected life of the equipment. In the case of inverters, the true cost of service for using central inverters is more easily computed than the cost for a string inverter system as the cost of the central inverter is usually known along with its related service contract, which is usually calculated from unit cost. However, with the string inverter which has no specific design life span and a relatively high failure rate, the cost to repair or replace, as become necessary, many units over a 25-30 years period, is much more difficult to compute, especially with the absence of significant warranties on string inverters. According to current industry data 1 string inverter per year will fail and by the 11th year a decision must be made whether to replace all existing units at one time or to continue replacing them as they fail only. As they fail the rate of failure is 3% in the first 10 years, 4% in the 11-15th year and 5% in the 16-20th year.

In general, the lifetime service cost for string inverters is higher than the same cost for central inverters and while string inverters clearly have a place in the utility scale inverter market some studies have shown that central inverters are far and away a superior choice for typical utility scale inverters in the US, all thing being equal. However, in locations where conditions are bad and there is an absence of required local skill sets, string inverters become a more viable option.

Finally, from the information shown above, central inverters have both CAPEX and OPEX advantage over string inverters over the life time of the system. According the Meywirth (2011) "A centralized system layout is a clear winner when it comes to ground-mounted large scale PV projects"

Benefits of Centralized Inverters

The major benefits of the centralized inverter are as follows:

- Simple Construction
- Easy to simulate through use of a model
- Low system cost
- Clean and dynamic system performance
- Cost effective communications network with only a few devices
- Higher control dynamics

- Future Viability with regard to grid management services.
- Proven technology for plants up to 700 MW

BIPV Systems

Silicon wafer and thin film PV technologies are usually utilized in the form of solar cells, modules and arrays all arranged in large fields or on rooftops to collect the incident solar radiation, however, this not the only way that solar cells, wafers or thin film, can be displayed. The solar cells, wafer or thin film, can be displayed as an integral part of several building components inclusive of the roof, skylights, façade, windows, railings and other exposed components of the building envelope. In this design the solar cells are integrated into the material that is utilized to form roofing material, façade material, the skylight, windows and other exposed areas of the building and the solar cells are affixed outward, facing the incident solar radiation in all directions. The use of BIPV systems provides many benefits to using the general PV systems to generate electrical energy with the main benefit being the financial benefits produced by eliminating the traditional support systems for traditional solar PV panels, the generation of electrical energy from solar systems without the need for additional and exclusive use of land, the utilization of vertical spaces on building facades that would not have been considered in the traditional systems and the elimination of the aesthetically unpleasant sight of solar systems installed on even the most beautiful of structures.

- BIPV systems could be utilized quite effectively in both rural and urban spaces, however, it would create far greater value in urban centers that usually have more buildings than open spaces and the building structures would automatically provide necessary support structures that would not be available in a rural setting.

CHAPTER 4

Power Transmission and Distribution Systems

In general, electrical power is usually generated at "power plants" in central locations, irrespective of the source of energy that is utilized to create the electricity, and the power generated at these plants are usually fed onto a national or regional grid from which users of all types are supplied. The basic customers of the grid usually include hospitals and healthcare facilities, schools and other educational facilities, residential facilities (domestic, commercial and institutional), commercial facilities, manufacturing facilities, industrial facilities and the security facilities inclusive of the police and the military and the requirements of all of these end users differ based on their demand for electricity. Electrical power is usually delivered in the following steps:

- Generation. The traditional methods of electrical power generations usually involved the utilization of a certain set of fuels inclusive of the fossil fuels, nuclear energy and hydropower plants the output of which are all connected to electromechanical generators which produce alternating current (AC) electrical power which can be converted to high or ultrahigh voltage electrical power in a transmission substation before it is delivered to a transmission system. In renewable energy systems electrical power is generated both as alternating and direct currents and the power generated from these systems are treated differently prior to being added to the transmission system. The AC power is sent to transformers and substations as previously indicated before being transferred to the electrical grid, while the DC power must be put through an additional step, inversion, before it become suitable for transformation and addition to the electrical grid.

- Transmission. The transmission system transports the stepped-up high voltage electricity power to the locations where it is desired and where it will be utilized.

- Distribution. Upon on arrival in the desired location the high voltage electricity is converted, stepped-down, to meet the lower voltage electrical power required by most users using distribution substations from which the required electrical power will flow to the customers.

Transmission

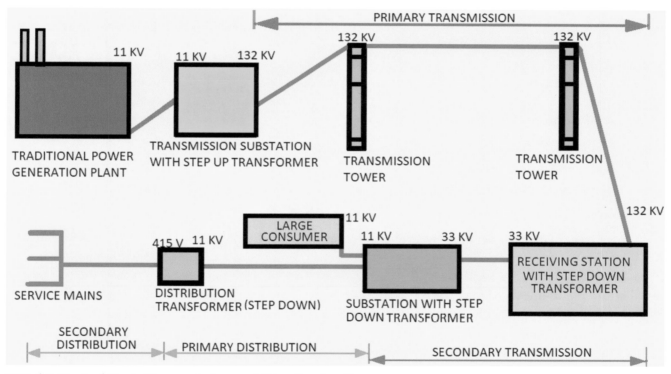

FIG 4.1 Typical Basic Transmission and Distribution System.

The basic function of the transmission system is the transportation of electrical energy over long distances, from the point of generation to the point where the electrical energy is to be utilized, with relatively low energy losses. The transportation of electrical energy is achieved using several components inclusive of conductors, wires and cables, support towers, transformers, protective devices, line insulators, and voltage regulators.

Conductors

The conductors are the most important element of the transmission system as they are directly responsible for the carrying and transportation of the electrical energy from one point to another, however, these conductors are not self-supporting, self-insulating, self-protecting and cannot make the necessary changes in voltage levels required by themselves and must therefore be supported by all the other major elements. The conductors to be used in transmission systems have certain basic requirement inclusive of high conductivity and low resistance, high current and voltage carrying capacity, high strength while being relative light in overall weight, pound per foot(kg/meter). From the chemical, mechanical and electrical conductivity properties two material have been found to be the best conductors for use in the transmissions of electricity and these materials are aluminum and copper, however, while they are the best conductors both materials have short comings that limits their use and they require the addition of other material to strengthen their capacities. Long experience and test data has shown that for high voltage and extra high voltages the following combination of materials are best suited for use in transmission systems:

- Hard or soft drawn copper conductors,
- Aluminum conductors,
- Aluminum alloy conductors,
- Aluminum conductor, steel reinforced
- Aluminum conductors, Aluminum-clad steel reinforced
- Aluminum conductors, Aluminum alloy reinforced

The conductor selected for service will be based on a number of factors inclusive of transmission voltage, installation requirements, project terrain conditions and the performance characteristics of available conductors. For high and extra high voltage service aluminum conductors are normally selected and the transmission conductor industry has developed a good selection of aluminum conductors that can be utilized in transmission, primary and secondary and distribution, primary and secondary and the conductors includes the All Aluminum Conductor (AAC) the All Aluminum Alloy Conductor (AAAC), the Aluminum Conductor Steel Reinforced (ACSR), the Aluminum Conductor, Aluminum Clad Steel Reinforced (ACSR/AW), Aluminum Conductor, Aluminum Alloy Reinforced (ACAR).

All Aluminum Conductor (AAC)

This is a stranded conductor that is made from aluminum with a purity of at least 99.7 % and it is most widely used in urban centers, where the spacing between supports are usually short, and in coastal regions due to it high resistance to corrosion. It has found wide application is overhead transmission and distribution systems with medium breaking loads, electrified wire networks, bus connections in substations and switch yards, and the solid conductors are used for mechanical and grounding applications.

All Aluminum Alloy Conductor (AAAC)

This AAAC conductor is a stranded Aluminum Alloy Conductor that has high mechanical strength and great corrosion resistance and these characteristics has allowed it to be utilized in many applications inclusive of usage at various voltage levels in bare overhead transmission and distribution systems, spanning across wide rivers and gorges, suitable for use in extremely cold icy locations and at any location that requires special engineering attention. Some other characteristics of the AAAC are longevity, large transmission capacity, large current carrying capacity, great wear resistance, great crush resistance, easy to install and maintain and a relatively low capital cost to design and install.

Aluminum Conductor Steel Reinforced (ACSR)

The ACSR conductor consist of a solid or stranded steel core surrounded by stranded aluminum with the steel core giving it greater strength which allows it to be used in applications requiring high strength conductors and less structural supports and such application would include spanning across wide rivers, gorges and mountain ranges. The major advantages of these conductors are their high tensile strength, light weight and the ability to modify the steel core without losing capacity, which allows these conductors to cover longer spans with fewer supports and these cables find wide applications in bare overhead conductors in transmission, distribution systems and overhead earth cables.

Aluminum Conductor, Aluminum Clad Steel Reinforced (ACSR/AW)

The ACSR/AW conductor consists of a core of stranded steel clad in Aluminum covered by strands of Aluminum which gives this conductor great tensile strength and a relatively light weight comparable to the ACSR, but the ACSR/AW IS lighter and stronger. This would make the ACSR/AW comparable to the ACSR in capacity and would allow it to be used in similar applications but at reduced material and installation costs. These conductors are quite suitable for use in bare overhead transmission and distribution systems, spanning rivers, gorges and mountains and in marine and industrial environments.

Aluminum Conductor, Aluminum Alloy Reinforced (ACAR)

The ACAR conductor consists of a stranded core of high strength Aluminum-Magnesium-Silicon alloy covered concentrically by strands of aluminum 1350 and the strength and capacity of this conductor may be determined by the number of strands in both the core and external layers. The ACAR conductor has better mechanical and electrical properties than the equivalent ACSR, AAC and AAAC conductors and this would make the ACAR the best selection where high ampacity, strength and light weight are the criteria properties for the design of bare overhead transmission and distribution systems. The ACAR conductor is suitable for utilization in overhead transmission, primary and secondary distribution systems and may be used in all applications requiring high strength, ampacity and light weight such as spanning wide spans with few supports.

Support Towers

The main function of the support tower is to bear the load of the overhead transmission conductors that are resting on it, maintain cables at a safe height above the ground, prevent the cables from ever touching and prevent excess movement in any direction during weather or seismic events and this requires that the towers must be structural sound and built on an engineered foundation. Support towers may be built from lattice steel frames, steel sections, wood or steel reinforced concrete and the tower normally consists of several sections inclusive of the peak, the cage, the cross arms, the boom, the body, the legs of the tower and the anchor bolts and baseplate of the tower. The figure below, FIG 4.2, highlights the different parts of the transmission tower.

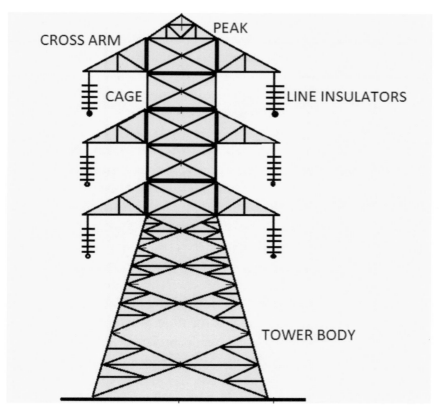

FIG. 4.2 Typical Transmission Tower.

Peak. The peak is the highest portion of the tower and it is above the highest cross arm as is shown in the figure above and the peak structure is used to support the lightening protection and ground conductor systems for the tower.

Cross Arm. The cross arms are the conductor carriers and they are deigned to support the weight of the conductors, insulators and accessories, keep the conductors away from the cage, the rest of tower body itself and from the other conductors in the vicinity. The structural strength of the cross arms is essential for bearing the weight of the conductors, related line insulators and accessories and its overall design will be a function of the voltage level of the power being transported.

Insulators. The most basic function of the insulators is to prevent the high or extra high voltage conductors from coming into contact with the towers and to ensure the complete isolation of the conductors to ensure the safety of the tower, the personnel working in or around it and the overall transmission operations.

Cage. The cage is that portion of the tower which supports the conductors, the insulators and the cross arms and must be structurally sound to ensure the longevity and the safe of operations of the tower.

Tower Body. The body is that portion of the tower that is in direct contact with the ground, terra firma, and it provides structural support for all of the other sections of the tower and its general operation and the upper part of the body also provides the minimum height of the tower and the minimum safe height of installing the high voltage conductors and it is very important that the body be designed to withstand all natural forces and any negative impact that the high voltage power may have on the material of the tower itself. The typical transmission tower can carry one or more circuits of high or extra high voltage transmission lines, with each circuit consisting of three conductors per circuit and each tower will have the ability to carry a minimum of three conductors up to as many as any one tower is designed to carry. The number of circuits per tower will be a function of the tower design, the amount of power to be provided to a particular locale and surrounding communities and the number and size of conductors required to deliver the electrical energy. The design of the tower will be a function of the electrical load, the number of conductors required, the current rating of the conductors and the level voltage to be transported, with higher voltages requiring higher towers to ensure the safety of facilities and humans that may come in close proximity to the path of the conductors. The most important factors that must be considered include the minimum ground clearance required, the length of insulator strings, the minimum clearance required between conductors and conductors and towers, the location of ground wire relative to the outermost conductors and the midspan clearance required, due to the dynamic behavior of the conductors and the lightening protection system.

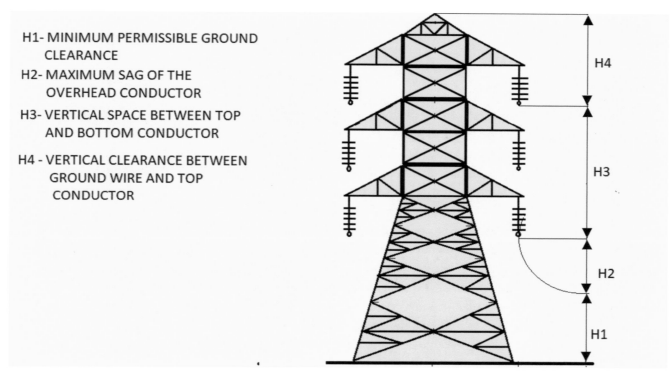

H1- MINIMUM PERMISSIBLE GROUND
    CLEARANCE

H2- MAXIMUM SAG OF THE
    OVERHEAD CONDUCTOR

H3- VERTICAL SPACE BETWEEN TOP
    AND BOTTOM CONDUCTOR

H4 - VERTICAL CLEARANCE BETWEEN
    GROUND WIRE AND TOP
    CONDUCTOR

FIG. 4.3. Typical Safe Height for the Installation of Conductors on a Transmission Tower.

Types of Transmission Towers

There are four types of transmission towers types A, B, C and D, each designed with different angles of deviation as follows:

- Type A – 0-2 degrees angle of deviation
- Type B – 2-15 degrees angle of deviation
- Type C – 15-30 degrees angle of deviation
- Type D – 30-60 degrees angle of deviation

The Type A towers may also be described as a tangential suspension tower while Type B, C and D may all be described as either Angle, Tension or Section Towers. These towers may also be categorized based on their specific

113

functions such as River Crossing, Rail or Highway Crossing and Transposition Tower or based on the number of circuits which they are designed to carry inclusive of One Circuit or Two Circuit or Multi Circuit towers.

Transformers

Alternating current (AC) transformers are the foundation upon which the cost effective and energy efficient high voltage transmission systems are built, as they give the electrical grid system the ability to increase the voltage for easy transmission and the ability to lower the voltage to allow it to be used by the general consumer and industrial users at the same time. The transformers used in the transmission system consist of two basic types, the step-up transformer and the step-down transformer both of which manipulate the relationship between power, voltage and current (power = voltage x current) to produce the desired increase or decrease in voltage. The standard transformer is made up of a laminated magnetic iron core and two sets of conductors that are wound around two different side of the magnetic core and the transformer operates when a voltage is applied to one side of the transformer, primary side, and this input voltage induces the setting up of a magnetic field in the magnetic core which then induces the flow of electricity in the output side of the transformer, secondary side. The relationship between input voltage and output voltage in a transformer are usually a function of the number of conductors turns and the size of the conductor on the primary and secondary sides of the transformer. In the standard transformer the conductors on the primary side are usually larger than the conductors on the secondary side and the number of conductors turns on both sides will be a function of the current capacity of the conductor and the desired output from the transformer. In practice the output side can have either a greater or lesser number of turns than the primary side of the transformer with the number of turns on the secondary side relative to the primary side determining when the output voltage will be higher or lower than the input voltage and whether the transformer can be described as a step-up or step-down transformer. The ratio of input conductor turns to output conductor turns will determine the ratio of the voltage change that will be achieved in the transformation process.

The change in voltage levels, up or down, does not affect the power relationship as input power is equal to output power, $P_i = P_o = V_i \times I_i = V_o \times I_o$

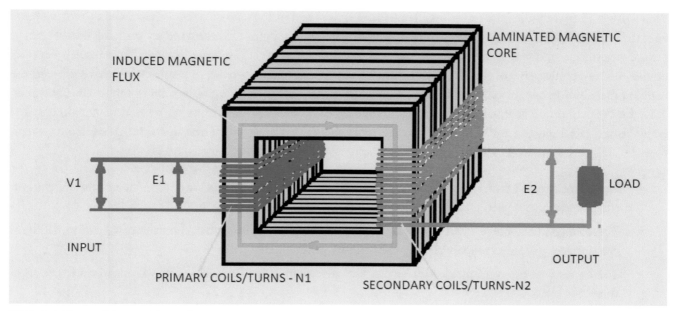

FIG 4.4 Typical Basic Transformer Circuits.

Step Up Transformer. The step-up transformer is usually located immediately after the generating plant to receive the output from the plant directly and it is usually located in a transmission substation which contains the transformer, safety switches, relays, lightening protection and other accessories to facilitate the safe and energy efficient change in voltage level from the generating plant to the desired transmission levels. The output of generating plants is usually in the region of 11 KV while the transmission levels are in the region of 33 KV, 66 KV, 132 KV, 200 KV up to 1100 KV and the only equipment capable of creating this very important change or transformation is the alternating current step-up transformer. Step-up transformer basics are as follows:

- The number of coils on the secondary/output side is higher than those on the primary input side.
- The voltage at output is several folds higher than that at input based on the turns-ratio, which is the number of secondary coils divided by the number of primary coils.
- The current at the output will be much smaller than at input and will also be a function of the turns-ratio.

Step Down Transformer. The step-down transformers are used in the transmission system when there is a need to transfer from higher voltage such as the primary transmission system to the secondary transmission system at a receiving station going from a voltage of 132 KV to a voltage of 66 KV, a lower voltage. The same also applies at all other levels inclusive of the step down from the secondary transmission to a substation that will feed the primary distribution system at 33KV and the steps down to secondary distribution system to supply the consumer, 11Kv for industrial customers, 460 Volts for commercial customers and 220 Volts for residential customers. The transformers at all stages are also to be supported by necessary switches, relays and lightening protection system. General step-down transformer basic are as follows:

- The number of coils on the secondary/output side of the transformer is lower that the number of coils on the primary/input side.
- The voltage at the output will be several folds lower that at the input based on the turns ration, which is the number primary coils divided by the number of secondary coils.
- The current on the output side will be much higher than at the input side and will also be a function of the turns-ratio.

Protection Requirements

The transmission system transport huge volumes of electrical energy on uninsulated conductors through the air over distances varying from hundreds to thousands of miles in many different environmental conditions, conditions that can negatively impact the continuous flow of electricity and also at many different levels. Some of these conditions include:

- Weather related events such as rain, wind, lightning, snow and ice
- Animals inclusive of birds, rodents and snakes,
- Trees
- Natural disasters inclusive of earthquakes, floods and fires
- Damage to transmission conductors and structures by aircrafts and automobiles.
- Faulty equipment inclusive of punctured, contaminated and broken insulators, damaged transformers, corroded and broken conductors and uncontrolled corona discharge.

Faults. A few of the typical faults that may develop from these conditions include:

- Single line-to ground (phase to earth)
- Line-to-line (Phase to phase)
- Line -to-line- to ground (Phase-to-phase-earth)
- Three phase power imbalances between phases.

Protection Systems. Several protection systems have been designed and built to minimize the many negative impacts that may arise from any event and these include:

- Overcurrent Protection Schemes that include LV line protection using fuses and circuit breakers and MV line protection using overcurrent relays connected to current transformers
- Differential and phase comparison schemes, used on short lines 80-100 km
- Distance protection, used on much longer lines
- Direction overcurrent
- Pilot protection schemes inclusive of Directional Comparison Blocking (DCB) and Permissive Over Reaching Transfer Trip (POTT)

Protection Devices. Some of the devices used in the protection schemes include the following:

- Protective relays
- Regulating Relays
- Reclosing and synchronizing relays – Auto reclose schemes
- Auxiliary Relays
- Single Pole Trips

All protection schemes are designed using sensitive devices that sense the possible problems and communicates these problems to a control device that reacts to prevent widespread damage to any section of the transmission systems.

Insulators

An insulator is a non-conducting material whose electron structure is static and does not permit the free movement of electron and includes material such as paper, plastic, rubber, glass and air. This capacity to prevent the movement of electrons allows insulators to be used in directing the flow of electrons in conducting materials, such as in electrical conductors and also in providing protection in the transmission of high voltage electrical energy by ensuring that the high voltage conductors are kept separate and by ensuring that the power carrying conductors never make direct contact with the metal frames of transmission towers and other metal support structures.

Types of insulators

There are several types of insulators which are used on transmission and distribution systems all over the world and includes discs insulators, post insulators, pin insulators, strain insulators, suspension insulators, shackle insulator, stay insulator, long rod insulator, polymer insulator and glass insulator.

Disc Insulators. These insulators which are shaped like discs can be made from either porcelain or glass and are highly utilized in standard high voltage, 11KV, transmission and distribution systems, usually on transmission towers and utility poles. They have very good insulation properties and high mechanical strength which are both essential to the successful transmission of high voltage electricity which is transported in heavy conductors that are supported in both tension and suspension structural systems. These insulators are of a robust design, have high corrosion resistance, have a relatively long life and are cost effective to use in most environments.

Post Insulators. These insulators are shaped like a cylindrical post and are made from ceramic or a composite, silicone and rubber, material that has very good insulating properties and high mechanical strength. These insulators are usually utilized in substations to protect transformers, in switchyards to protect switchgear and other connecting equipment which utilize extra high voltages up to 1100 KV. The post insulator is generally light weight and the combination of very good insulating properties, high mechanical strength, very good chemical and thermal strengths gives it a low risk of damage and a relatively long life.

Pin Insulator. These insulators are usually made from porcelain or glass and are highly utilized in standard high voltage systems, 11 KV, distribution systems, on utility distribution poles. These insulators are designed to separate the high voltage conductor from the metal pins that attaches the insulator carrying the conductor to the cross arm of the utility poles and they have high insulating properties and high mechanical strength.

Strain Insulator. These insulators are generally made from porcelain, glass or fiber glass and are designed to operate under stress while supporting different elements inclusive of radio antennas and overhead conductors. They are utilized in two main areas, connecting conductors to towers or poles, utilizing their insulating capacities and ability to withstand tensions and separating high voltage, 33 KV, conductors, utilizing their strength and insulating capacities.

Suspension Insulator. These insulators are generally made from porcelain and consist of a number of insulators connected in series and suspended together on a string from which the conductor is carried on the lower end below the insulators with the upper end of the insulator attached to the cross arm of a transmission tower. These insulators have insulating capacities and high mechanical strength to support the high voltage, 33 KV, conductors that the carry on their lower end.

Shackle Insulator. These insulators are generally made from porcelain, are usually small in size relative to other insulators and are used in secondary distribution system carrying voltages up to 33 KV. They are fairly versatile as they can be utilized in both a vertical and horizontal position as required and they have very good insulating properties and high mechanical strength.

Stay Insulators. These insulators are made from porcelain, are relatively small, are rectangular in shape and are generally utilized on low voltage systems such as being part of guy wires, that support dead end poles, by fastening them to the ground and providing necessary counter weight.

Long Rod Insulators. These insulators are made from porcelain rods with weather sheds and metal ends on the outside and consists of a combination of multiple insulators that are utilized on metal transmission towers to separate the conductors from the tower itself. They are fairly reliable as they rarely breakdown, have long life spans, provide better insulation properties under polluted conditions and low weight with simple mounting systems.

Polymer Insulators. These insulators are made of a combination of fiberglass, polymer and metal fittings and usually consist of a fiberglass core surrounded by polymer weather sheds that shield the insulator core from the external environment. These insulators are much lighter in weight than those made from porcelain but provides much better insulating properties and mechanical strength.

Glass Insulators. These insulators are made from toughened or annealed glass and are normally used on high voltage transmission lines to prevent the leakage of electricity into the towers and poles and then into the ground, which is a significant safety issue. The glass insulators have significant advantages over the porcelain equivalent as they have higher die electric strength, higher resistivity and a lower coefficient of thermal expansion.

Voltage Regulators

Voltage regulation is critical for the delivery of power at a constant or near constant voltage level at all times irrespective of what may be happening on the supply or load side of the power delivery scheme. This activity, voltage regulation, may done by a dedicated voltage regulator or by a transformer that was designed to automatically or manually regulate the voltage to ensure constant delivery. These transformers are designed with the ability to modify the transformer take off points or taps to achieve changes in voltage delivery and are referred to as On Load Tap Changer (OLTC), which automatically make the changes or Off Load Tap Changer (OLTC), which must be done manually with the transformer out of service. A stand-alone voltage regulator working in conjunction with a transformer without the tap changing ability can similarly provide this service and perform them at a lower cost and quite effectively. These voltage regulators are designed to be utilized directly with the transformers, step-up or step-down, in substations in both transmission and distribution substations, on distribution poles or other installations where transformation service is required.

The transmission voltage regulator has one main function, provide and maintain a constant voltage output to the transmission lines and to transmission customers, even when problems arise with inconsistent supplies from the power plants, from other areas of the distribution systems and the other technical issues that may be affecting the transmission systems due to distributed generation(DG) power systems and the intermittent supply of electrical power from some renewable energy plants inclusive of wind and solar systems.

Distribution Systems

The distribution system is that part of the electrical system between the transmission substation, that is fed by the transmission system, and the customers and this distribution system usually consist of feeders, distributors and service mains. Feeder cables are designed to have a greater load carrying capacity than both the distributors and the service mains, while a distributor will have a greater capacity than a service main.

Feeders. A feeder line is connected directly to the substation and supplies electrical power to distributors only, that is no customer is directly connected to a feeder line as is shown in the diagram below.

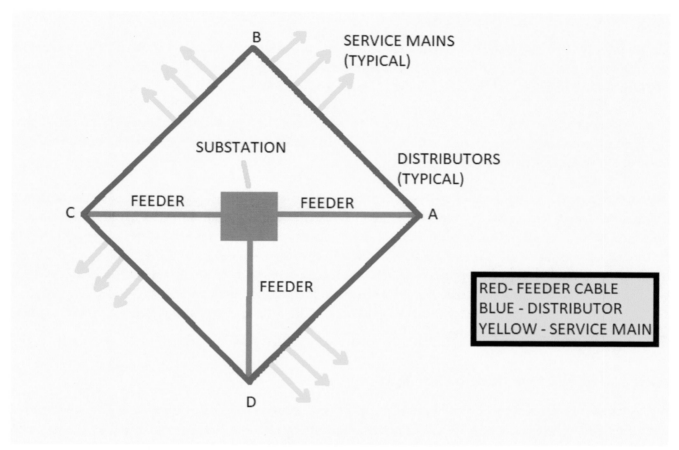

Fig. 4.5 Typical Arrangement of Feeders, Distributors and Service Mains.

Distributors. The distributor is the conductor that is designed to supply customers with required electrical power and to which the conductors connecting the customer to the power distribution system is attached and the size and length of these distributors are a function of the number of customers to be served in a particular area and regulations covering the allowed voltage drop in distributors.

Service Mains. The service mains are relatively small cables that provide the customers a direct link to the power supply and it is usually attached to a distributor as highlighted in Fig 4.1 and the size and length of this cable are a function of the power requirements of the customer and the regulation governing the permissible voltage drop in the service main.

Classification of Distribution Systems

Distribution systems may be classified based on several factors inclusive of the nature of the current, alternating or direct current, that is to be distributed, the type of construction used to establish the required infrastructure of the distribution system and the scheme of the connections to be used.

Nature of the Current. The nature of the current is a function of the type of electrical power being supplied to the customer, alternating current, AC, or direct current, DC, and the requirement for the existing electrical grid is AC and therefore the focus shall be on AC.

Type of Construction. The type of construction refers to the structural support system used to install necessary distribution cables and includes overhead supports systems and underground systems. Both overhead and underground systems are well established and widely used, however, overhead systems are utilized in probably 90 percent or more of all distribution systems worldwide and the reason for this great difference is the capital cost to establish the different system, as the cost to create the underground system is five to ten times greater than that required to establish an equivalent overhead system. Underground systems are mostly used where overhead systems are impractical, where local laws and regulations prohibit the use of overhead systems or when customers find them more aesthetically pleasing.

Alternating Current Distribution Systems

In general, the electrical energy provided by most traditional power plants is generated, transmitted and distributed as alternating current, however, some renewable energy electrical power generation plants generate only direct current which must first be converted to alternating current before it can be added to the transmission system or grid for transmission and later distribution. Electrical power generation, transmission and distribution commenced using direct current only, however, the development of alternating current a little later saw the decline of direct current system and the rise in the almost exclusive use of alternating current. Alternating Current, AC, became more popular due to the following:

- Alternating voltages can be conveniently raised or lowered to meet almost any need using AC transformers.
- High AC transmission and distribution voltages allow for greatly reduced associated currents, based on $P = VI$, when the power is constant the current will decrease as the voltage increases. The reduction in transmission current will allow the utilization of cables with smaller cross-sectional areas which greatly reduces the weight of the required cables and this reduction in weight will impact the structural design and spacing of the transmission.

This reduction in the sizes of the transmission cables will also help to reduce power line losses, reduces the environmental impact of the transmission systems and also reduces the capital cost to construct the systems.

- AC systems are used in all phases of providing electrical energy to the public, inclusive of the transfer of energy to the transmission system, the transportation of energy to the location where it is to be utilized, the transfer of energy to the substation, the transfer to the local distribution system and finally the transfer to the point of use.

Primary and Secondary Distribution Systems

The primary distribution system focuses on the delivery of electrical energy from the transmission system to large users and to the secondary distribution system.

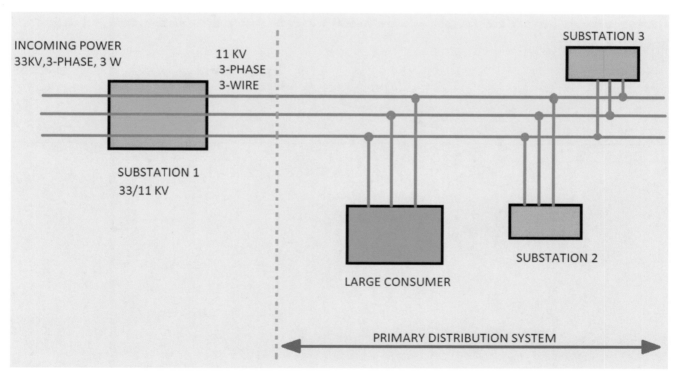

FIG. 4.6 Typical Primary Distribution Systems.

The primary distribution system is that part of the electrical distribution system that usually operates at much higher voltages than the voltages utilized by the general consumer and these higher voltages are usually in the region of 11 KV, 6.6 KV and 3.3 KV and are usually focused on the large consumers.

Secondary Distribution Systems

The secondary distribution system is that portion of the electrical distribution system which supplies electrical energy at the 400 V/3-Phase and 220/115 V 1-Phase levels and the primary distribution system supplies power to secondary distribution substations which are located in close proximity to the consumers and these substations contain step down transformers, relays, switches and protection devices all of which were specifically designed to safely meet the consumers demand for electrical energy. At each of the secondary distribution substation the higher voltage is lowered from 11 KV to 400 V and the energy is delivered by a 3-phase, 4- wire system. In this system the voltage between any two phases will give 400 V and the voltage between any phase and neutral will give 220 V/240 V

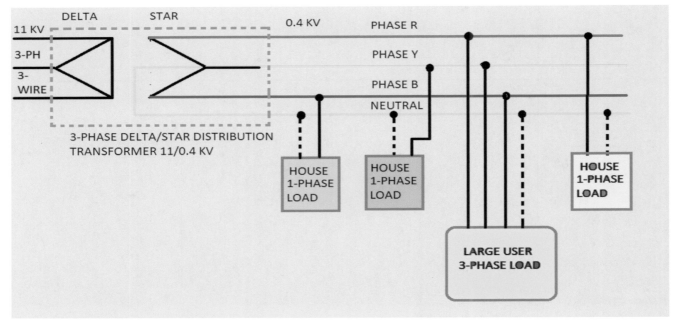

FIG. 4.7 Typical Secondary Distribution System.

Advantages of the AC System

There are many advantages to be gained by using the AC system inclusive of:

- Power may be generated at high voltages.
- The maintenance of AC substations is relatively easy, with relatively low cost.
- AC voltages may be stepped up or down using transformers quite easily and efficiently and this ability allows for the generation and transmission of electrical energy at very high voltages and its safe utilization at much lower voltages.

Disadvantages of the AC System

- AC lines and cables require more copper
- The construction of the AC distribution system is much more complicated.
- Skin effect in an AC system causes an effective increase in line resistance.
- AC lines have line capacitance which causes a continuous loss of power due to charging currents even when there is a no-load condition.

Requirements of an Electrical Distribution System

The most basic requirements of an electrical distribution system are proper voltage, availability of power on demand and reliability.

Proper Voltage. One of the most important requirements for an electrical distribution system is the consistent delivery of electricity of a particular voltage, current and frequency all the time, twenty four hours per day, seven days per week and fifty two weeks per years ad infinitum to the consumers take off point, as any change or frequent changes in any of these parameter will have negative consequences for the consumer as high or low voltage, high or low currents and high or low frequencies may cause significant damages to the consumers property, this is especially true for voltages. Typically, any variations in loads, especially from large consumers, on the distribution system can cause significant variations in voltages on the distribution system, either high or low voltage, can destroy lamps and motors in appliances in most domestic appliances and also most motors used in industry.

The delivery of electrical energy on both the transmission and distribution systems are always impacted by operating conditions and outside events which are not in the control of operations management and these impacts will cause fluctuations in voltage that will be delivered to the consumers terminals or take off points. These fluctuations, per established regulations, must be controlled within certain limits and per the regulations the allowable limit of variation in electricity supply must be within the range of plus or minus six percent of the rated voltage at the consumers terminals.

Availability of Power on Demand. Electrical energy must be available at all times irrespective of the time of day and the number of consumers, large and small, that may be desirous of using electrical energy at the same time. At periods of high demand when everyone needs to use electricity the supplier must devise strategies that will allow them to effectively meet all the demands and the strategies used to accomplish this may include redundant equipment that is brought on line to meet high demands or schemes that will facilitate a schedule change in some consumers demand for electrical energy. These schemes are usually referred to as demand side management as the supplier works with the consumer to manage the timing of their demand for electrical energy in a manner that will benefit all.

Reliability. The profitability and safe operations of all facilities inclusive of homes and businesses require that the energy that they need to operate be available on a consistent basis at all times as this will allow them to effectively plan their daily activities and to utilize their resources for their maximum benefit. Unfortunately, no system, inclusive of power generation, transmission and distribution, is one hundred percent reliable and suppliers must plan and create systems that can be depended on to improve the reliability of their operation. These plans could involve the installation and utilization of redundant or back up equipment, the installation and utilization of interconnected systems and the installation and utilization of reliable automatic control systems.

Schemes of Connection

There are basically three schemes of connections that are used in most alternating current distribution systems and these schemes include the radial system, the ring main system and the inter- connected system.

Radial System. In a Radial System the feeders radiate from a substation and feed the distributor from one end only as shown in the diagram below.

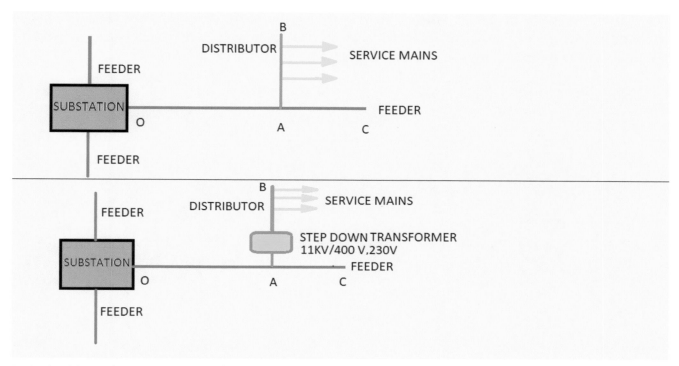

FIG. 4.8 Typical Line Diagram of a Radial Distribution System

The radial system is the least complicated of the connections schemes and it can be designed and constructed at the lowest capital cost, relative to the other systems.

Disadvantages of the Radial System.

- The point of the distributor closest to the feeder, connection point, will be most heavily loaded.

- All failures on the feeder and distributor will negatively impact all customers who are on the side of the fault.

- Consumers located at or near the end of the distributor are usually subjected to voltage fluctuations when there are any significant changes in loads.

These limitations ensure that radial systems are only suitable for relatively short cable runs only.

Ring Main System. The Ring Main System is based on the loop formed by the primary side of the distribution transformers and begins at the substation bus bars followed by a loop through the community to be supplied with electrical energy before returning to the substation as demonstrated in the figure below.

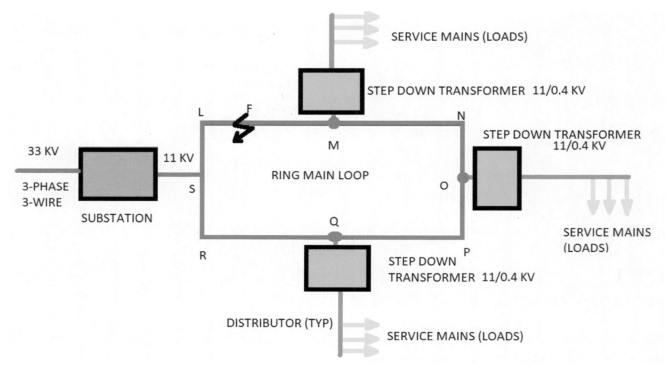

FIG. 4.9 Typical Line Diagram of Ring Main Distribution System

Advantages of the Ring Main System

The main advantages of the ring main system are as follows:

- Customers experience less voltage fluctuations
- The ring main system is very reliable as each distributor is fed by two feeders which ensure continuity of service if there is a failure in one feeder

Interconnected System. An interconnected system is created when a feeder loop is fed by two or more substations or generation plants. The diagram below highlights the interconnected system as shown in FIG 4.10 below.

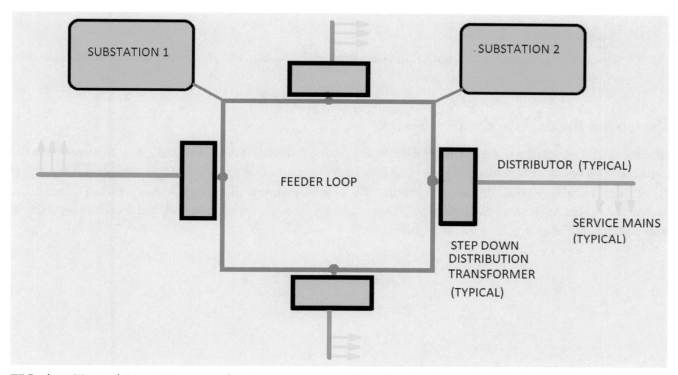

FIG. 4.10 Typical Line Diagram of an Interconnected Distribution System.

Advantages of the Interconnected System

The interconnected system has several significant advantages inclusive of the following:

- Increased service reliability
- All locations fed by one substation or generating plant during peak hours can also be supplied by the other substation or generating plant.

Protection for the Distribution System

Similar to the transmission and primary distribution system, the secondary distribution system must also be provided with voltage regulators and protection systems to ensure that the proper voltage, current and frequency is delivered to end users at all times and the protection schemes and devices utilized in the secondary distribution system are similar to those utilized elsewhere, however, the devices are specifically designed for the lower voltages and currents that is normally supplied to end users.

CHAPTER 5

Solar Energy Grid Integration System (SEGIS)

The growth of grid connected renewable energy systems inclusive of residential, commercial and utility scale CSP, PV systems, wind energy farms and nonrenewable distributed generation systems, such as small combined heat and power plants (CHP), will all create a significant impact upon the established traditional electrical transmission and distribution systems due to the distributed and intermittent nature of some of these new sources of electrical power being added to the grid. Prior to the advent of alternative renewable sources of energy and other distributed generation systems, the grid was fed by large central electromechanical power plants which utilized several different sources of energy inclusive of oil, coal, natural gas, nuclear and hydropower and these power plants were located at relatively significant distances from each other and their points of connection to the electrical grid. The combination of new electrical energy sources from many small to large size power plants in many different locations is usually referred to as distributed generation and according to Purchala et al, the IEEE defines distributed generation as "the generating of electricity by facilities that are sufficiently smaller than central generating plants so as to allow interconnection at nearly any point in a power system".

Impact of Distributed Generation (DG) on Grid

The development of distributed generation has been fostered by five important factors according to the International Energy Agency (IEA) and these factors include:

- The development of renewable DG technologies inclusive of Solar, Wind, Bioenergy, Geothermal and mini-hydropower systems.
- The cost constraint related to the construction of new major transmission lines
- The increased customer demand for highly reliable electricity

- The liberalization of electricity supply market.
- Increased concerns about the effects of Climate Change

The development and utilization of DG systems while quite beneficial overall also come with some negative impacts that must be mitigated in order to obtain optimum output from the systems. The main benefits provided by the DG systems are as follows:

- Helps countries and communities to meet environmental goals inclusive of the reduction in the production and emission of greenhouse gases in the production of needed electrical energy, which are responsible for climate change.
- The development and utilization of DG systems create significant economic activity and will help to create a green economy.
- Reduces dependency on the purchase and importation of fuels from external source in an energy market that can be quite volatile, thereby increasing national energy security.
- Significantly improving electrical energy supply reliability and quality.
- Reduces the need for new grid investments.
- Make the electrical supply more flexible to meet price response, reliability and quality needs.
- Reduces grid transmission and distribution losses.
- Improves grid system security, efficiency and reliability.

The constraints associated with the development and utilization of DG systems are as follows:

- Financial costs. The development and construction of DG plants have a higher cost per Kw installed compared to large central power plants.
- Predictability of power output. It is difficult to predict power output from some renewable energy plants, especially solar and wind energy systems that are directly dependent on the changing weather conditions and the time of the day.
- Power Quality. Voltage and frequency level control issues and reversal of power flow.
- Network protection issues. These problems include false tripping of feeders, blinding of protection systems, increased fault levels, unintentional islanding, unsynchronized reclosing and prevention of automatic reclosing.

- Connection issues. These include the need for different protection schemes for the different voltage levels due to bidirectional power flow, anti-islanding protection and resynchronization requirements.

- Generally, while there are some negative impacts related to the utilization of DG systems, these systems help countries to meet their environmental goals, provide significant economic benefits, improves energy security and improves electrical power supply reliability. The net positive outcomes from using DG systems therefore require that the necessary technologies be developed and utilized to facilitate their smooth integration onto the electrical grid and their smooth operation alongside existing older power generation technology.

PV System Grid Integration

The growth in the use of distributed generations systems and their connection to the electric grid within the last 30 plus years has created the need for more sophisticated operating and control systems to better manage the connection of these systems to the existing electrical grid. Currently most residential and small commercial PV systems are connected to the grid using a net-metering, grid connected approach, which while sufficient for previous iterations of PV energy systems needs to be updated to include other approaches such as the time-of-use and peak demand rate structures which require more sophisticated systems that are designed to integrate energy management and energy storage systems into the total PV systems architecture. The integration of energy management and energy storage systems would automatically change the architecture of the existing system and this updated architecture would include ways for controlling power flow into and away from the grid, systems to ensure grid reliability, power quality and alternative protection systems to accommodate a large number of distributed generation sources. The required changes in systems architecture would greatly improve systems operation and as a result accommodate a greater penetration of PV systems generated electrical energy into the electrical grid.

In order to fully understand the nature of the problem and the required need for changes we need to examine how existing renewable energy and distributed generations system work and how they impact the electrical grid. Some of the problems are as indicated below

a. Existing PV systems have the following characteristics when the PV Array and inverter are connected in parallel to the load:
   - The load is served only when the grid is available.

- Electrical energy produced by the PV systems tend to decrease the apparent load and the electrical energy produced that is greater than the demands of the load, flows into the grid.

- The PV system has no backup energy storage system and cannot serve the load in the absence of the grid.

- PV systems only produce electrical power with a power factor of one and any required volt ampere reactive (VAR) power must be provided by the grid.

b. Existing inverters must meet the requirements of the IEEE 1547- 2005 and therefore have the following characteristics:

- There is no direct communications or controls between the utility and the inverter

- When the inverter senses that the utility service has fallen outside of set boundaries for voltage and/ or frequency or there is an interruption in the utility service the inverter will disconnect from the utility until regular service condition has been restored, while the load remains connected to the utility.

c. Existing residential and small commercial PV systems connected to the grid are net-metered at a flat rate and have the following properties:

- The daily price of electrical energy is constant and there are no demand charges.

- The production of excess energy above load requirements causes the meter to spin backwards

- Electrical energy is bought and sold at the same rate.

- If energy produced exceeds energy used, over a period of a month or a year, the utility will not make payments for the excess above the amount used.

- When the utility is not available, grid-tied PV systems, without energy storage and load transfer capabilities, cannot serve the load even in the presence of bright sunlight and with the modules ready and able to create electrical power.

d. Large scale commercial and utility scale PV and other renewable systems normally operate on more complex rate structures some of which are as follows:

- Time-of-use rate. With these rates the cost of energy is usually higher during periods of peak demand.

- Demand charges apply when a significant portion of the utility bill is derived from the highest power requirements (KW), measured over a 15-30 minutes interval during the monthly billing period.

- Separate charge for VARs may be applied.

- Net-metering is less common and some systems are not permitted to deliver any power to the utility grid and in such a system the load must always exceed the capacity of the PV system.

- Some systems may have dual metering and in these systems the utility will purchase the power supplied by the PV system at a lower rate than the power supplied by the utility.

Problems for Utility Operations

There are some general concerns that the high penetration of inverter-based PV energy systems will contribute to instability and possible unsafe operating conditions on the distribution system due to several design and operational problems, inclusive of the following:

a. PV power production does not always coincide with the times when it is most economical for utilities to utilize it and this can negatively impact the economics of operations in the following scenarios:
   - The unavailability of electrical energy from PV systems in the hours immediately after sunset when demand for power is normally high, forces the utilities to increase power generation utilizing fossil fuels to meet peak demands during these hours at a high cost.

   - Utility demands are low during the early morning hours, sunrise to 9:00 am, and power from the PV systems during this period results in a lower load for the utility and there is a decreased need for an economical 24-hour base load power and this increases the need for more expensive intermediate and peaking power for the rest of the day.

b. From the utility operator's perspective, the net-metered flat rate customers do not pay their fair share of the operating cost and this more so with customers whose net demand is very close to zero as explained below:
   - When energy generated is equal to energy used, the energy related charges will be close zero and energy related charges is a large portion of most residential and small commercial power bills.

   - Without a demand charge or significant interconnection fee small customers usually pay little or nothing for the benefit of being connected to the grid.

c. The production of electrical energy from one PV system will increase or decrease quite quickly based on meteorological conditions such as clouds passing overhead.
   - In the majority of cases, the rate of change of the collective output from the PV system will be moderate, however, when the system is relatively small rapidly-passing cloud banks may eliminate all solar generation in less than five minutes (Tucson Electric Power).

- The injection of a significant amount of rapidly changing electrical power into a grid system may cause instability and affect the controls and increase the need for a spinning reserve.

d. When a utility experience sagging voltages under high demand conditions, the IEEE 1547 requires that inverters disconnect, but since the loads are not automatically disconnected, the utility will see an increase in demand, which will potentially aggravate the cause of the voltage sag which may lead to a power outage, decreasing the utility system reliability.

e. The addition to the grid of a large number of inverters, with increased PV penetration, has been shown to increase the probability of problems such as islanding where the inverters continue to supply power to local loads after a utility outage, with several possible negative impacts as shown below:

  - Existing inverters are usually limited in their ability to create extremely high levels of short-circuit currents, however, the addition of larger inverter systems or many small inverter systems can sum to significant short-circuit currents that could possibly cause serious equipment malfunction or damage.

  - Most utility protection systems are designed to detect a fault such as an arc-to- ground caused by tree branches falling across lines, in which case a relay briefly opens disconnecting the system from the fault to allow the fault to clear, after which it then reconnects the system to provide continuity of service. However, if the islanding detection fails and the inverter remains online, the inverter may be damage when the system is reconnected or the inverter may continue to supply power which could maintain the fault and consequently would cause the utility protection system to lock open. Such a condition would require the use of utility technician to correct the fault or reset the relays, and consequently may cause customers to be without power for a significantly longer period of time. Secondly, if the inverters are still supplying power to the grid, the technicians could be exposed to significant safety problems due to downed power lines and power to the load-side of the disconnect.

f. Large power flows into distribution systems that were designed for one-way flows may impact the ability of the system to regulate voltages and to provide protection and while the full impact of deep PV penetration is not yet clear there are several issues that have been identified, inclusive of the following:

  - Reverse power flow affects voltage regulation. Voltage regulators measure current and voltage and will maintain higher voltages at the beginning of a radial line in proportion to the power flow, however, the introduction of significant distributed power downstream of the voltage regulation system will cause line loading to appear low.

  - If the power is injected just downstream of the voltage regulators, customers at the end of the line will experience low voltage.

  - If the power is injected near the end of the line high voltages may occur at that point.

- Electrical protection fuses are designed to protect the current carrying capability of the line, the injection of power downstream from a fuse will not be detected and this presents the potential for a power overload.

- Frequency regulation may also be affected by the injection of distributed power especially when the distributed PV power becomes greater than the local conventional power generation system.

- It has been observed that phase-to-neutral overvoltage may develop with load to generation imbalance or phase-to-neutral faults and that this condition may become worse with high penetration of distributed PV systems with no dispatchability or interactive controls. The disproportionate installation of single- phase systems on a single-phase line may cause severely unbalanced networks which could lead to severely damaged controls and or transformers.

- Inverters using Pulse-Width- Modulation (PWM) schemes to regulate their output typically do not add to lower number harmonics, but the higher frequencies associated with power electronics will inject higher order harmonics.

- The levels of high order harmonics from power electronics in inverters could interfere with distribution systems inclusive of control and protection system, however, this problem may be mitigated by the addition of costly filtering systems.

- Inverter-generated pulses associated with impedance detection, for anti-islanding systems, will also accumulate in high penetration scenarios and cause out-of-spec utility voltage profiles.

Each of the problems highlighted above has the capacity to decrease system reliability or to severely increase utility costs in high penetration distributed PV scenarios.

Implications for Solar Systems Owners

All of the problems identified will impact the utility systems and the owners of the PV systems in significantly different ways, for the owners, some of the possible outcomes are alternate metering strategies and or limits on the amount of power that they can supply to the grid. Some of the alternate metering strategies are as follows:

- Net-metering may be replaced by dual meters which would allow the utility to apply one rate for electrical energy supplied by the utility to the customer and a lower rate for the electrical energy provided by the customer. Alternatively net-metering may be replaced by meters which turn in one direction only, which would give no credit for the power supplied by the customer to the grid.

- Time-of-use rates may be imposed to reflect the true cost of the utility supplying electrical energy to balance the high rates for electrical energy from sunset to 9:00 pm, peak demand, which significantly reduces the value of the PV system as the demands for the PV systems decline sharply.

- The utility rate structure may be changed by lowering energy charges and imposing infrastructure and demand charges that capture the true cost input of peaking power penetration and peak loads on the distribution system.

- Businesses and industrial sites with large motors could be provided with VAR power by the utility with necessary charges applied. Such a solution would be advantageous until renewable energy systems can produce the necessary VAR.

Approaches to Enable High Penetration

Today's distribution systems have relatively long useful lives ranging from 30 years design life to much longer periods of time and existing PV system must be designed to meet the design requirements of the distribution systems. With deep penetration the requirements will remain the same, however, the impact on the distribution system will at best be uncertain until new design criteria for the distribution and PV systems are developed to make them more compatible to facilitate greater PV penetration.

Another option to making the distribution system more reliable is to force the owners of new distributed energy systems, PV, CSP, wind, CHPs and mini water turbines, to pay for fixing the distribution systems to facilitate the injection of the energy that they produce into the existing electrical grid, however, the cost for the work required would be too onerous for small energy producers and would make their operation uneconomical. The more feasible approach to deeper penetration should include mitigating the negative impacts on the current distribution infrastructure and improving value for the solar energy system owners.

Mitigating Impacts on Electrical Power Distribution Infrastructure

There are many steps that can be taken to mitigate the impact of high DG penetration without completely replacing the existing grid infrastructure and these steps assume that the grid remains unchanged except for major changes in communication capacity of the grid system. The first step in improving the communication system is to allow the distribution system to interactively control the operations of the inverter/controller and one possible approach is to replace the existing anti-islanding control system with interactive dispatch that will enable the utility to command an inverter to ride through voltage sags, instead of having a large number of inverters go off-line while leaving the load connected (Sandia Laboratories). Interactive controls will also enable the distribution system to direct the inverter to go off-line whenever a fault arises instead of relying on multiple inverters to independently detect a fault or islanding condition.

Additional impact mitigating steps could involve enabling the management of power flows to and from the grid, this management process would keep high electrical energy flows off the distribution system when these high electrical energy flows could interfere with voltage and frequency regulation and protection systems, to provide power flows when the power is needed and available and to mitigate rapid transients as necessary. In general, the ideal loads for a centralized power plant should be predictable and constant or should be large when the system demand is low and small when system demands are high. A central control system could be utilized to manage the electrical energy flows by directing loads to operate when solar energy is available, to not operate during peak demand periods when solar energy is not available and to dampen the impact of transients. Some of the other steps are as follows:

- The addition of an energy storage system that would provide more flexibility to load management.
- The modification of the inverter or inverter-controller to allow them to produce VARs and assist with voltage regulations by producing VARs under the direction of the distribution system and adding power dispatchability.
- The addition of communications capability to allow the distribution system to signal the control system to dispatch electrical energy and loads to optimize power flow to the utility. These communications shall be in the form of time-of-use rates, demand charges or real time pricing by the way of smart-metering technology.

Improving Value for Solar Energy Systems Customers/Owners

All modifications or improvements made to facilitate the deeper penetration of PV and other distributed generation systems will not only impact the operation of the distribution systems and the owners of the distribution system, but these changes will also impact the customers or owners of the distributed generation systems. The customers will be impacted by the transition to time-of-use rates and demand charges as their compensation for energy delivered to the utility will be reduced at particular times of the day and their bills for energy received from the utility will be higher at the same time of day. The impact on the customer can be minimized with the use of a solar energy grid integration system (SEGIS) (Sandia Laboratories) which employs control strategies that optimize value. The SEGIS could mitigate the effects of the new rate structures or it can improve the economic position of the customer by providing values much higher than they currently receive. The SEGIS would be used to manage the flow of power to and from the utility so that power is purchased from the utility mainly when rates are low, peak demand is minimized and power would be sold to the utility when rates are high.

- Optimizing value for the customer would first take into consideration the orientation of the PV modules, azimuth and tilt angles, and the proposed rate structure and in regions of the world where utility demands peak in summer and time-of-use rates are utilized, designing the module array with one portion of the array facing west could provide the customer with more value, even though the total energy delivered may be lower.

- The second step involves dispatching the load to operate in concert with the availability of solar energy and or inexpensive electrical energy from the utility. The utilization of direct communication between the inverter and smart load supported by standard communication protocol, or by communication between the inverter and an energy management system could produce the desired result.

- The third step in this process would be the addition of an energy storage system to facilitate the capture of excess solar energy or inexpensive electrical energy from the utility that can be utilized at a later point in time when there is a great demand, and solar energy and inexpensive electrical energy from the utility is not available. These energy storage systems provide two basic services, improving the value of the PV system and enabling building critical systems to operate when there is a utility outage.

The optimization of the PV energy usage without undue burden on the owner and or operator typically require the utilization of a control system with adoptive logic and such a controller would monitor demand, the availability of solar energy supply and utility rates, all of which would be utilized to optimize the flow of energy based on time of day, day of the week and time of year, by controlling dispatchable loads and storage operations. The controller could also communicate with the utility smart-metering devices to obtain real-time pricing and monitor weather trends and forecasts, by way of the internet, to anticipate the availability of solar energy resources and the real-time price of solar power.

Advanced Distribution Systems and Microgrids

There is today a greater need for the utilization of renewable energy resources in order to minimize the use of fossil fuels in the efforts to mitigate against the many negative impacts of the carbon economy, which is based on fossil fuels, due to the negative impacts on the natural environment, and these negative impacts include the increase of carbon dioxide within our atmosphere, which consequently leads to the warming of planet Earth and the phenomena of climate change. In order to increase the amount of renewable energy used, specifically solar and wind, in the energy mix, new energy distribution systems must be created or old systems modified to facilitate greater usage of energy created by these sources in a manner that will not totally disrupt existing systems of energy supplies to customers internationally.

According to the USEIA (2018) solar and wind energy currently supplies only about 1% of the energy needs of peoples around the world and if these energy resources are to have a greater impact these levels of availability and usage must be increased to 30-40% and this greater utilization of renewable energy would require greater access to the electrical grids around the world, generally greater electrical grid penetration. The existing grids therefore need to evolve over time to produce a more advanced distribution system that will accommodate two-way power flows that will better facilitate and utilize distributed energy resources. To achieve higher levels of renewable energy usage and penetration, the utility companies should utilize advanced distribution and mini-grid systems combined with smart loads, all of which will help them to produce and deliver clean, constant and reliable electrical energy to their consumers while significantly reducing their negative impact on the environment. The approach to be taken by the utilities could include the use of three specific systems which are as follows:

- Placing utility owned distributed generation systems within the existing distribution system to improve operations and reliability,

- Placing energy storage systems within the existing distribution systems to improve the value of their PV system and improving the reliability of operations.

- The utilization of mini-grids. These mini-grids when placed at critical weak points in the distribution system can help to stabilize the grid and keep loads connected to it fully supplied even during a utility outage.

Smart Grid

An intelligent interactive electrical grid can enhance the safety, efficiency of production and utilization of electrical energy produced from many different locations, while using different energy resource within every nation. The need for such a system has become even greater with the advent of Climate Change caused by anthropogenic activities, basically the presence of excess carbon dioxide within our atmosphere caused by the amplified usage of fossil fuels since the start of the industrial revolution, and the need to utilized alternative non- carbon producing renewable energy sources in the process of minimizing and or eliminating the need for and the use of these fossil fuels. The existing electrical grid, its protection systems and many accessories were designed and built for the existing power plants which promotes a unidirectional and a constant steady flow of electricity from supply source to end users. However, to achieve the desired goals of reducing carbon emissions into the atmosphere the architecture of power production must be changed from a central system to a more distributed one utilizing every space and source of energy available.

The distributed production of electrical energy from different energy source, mostly wind and solar, is very different from the historical power generation systems and presents a set of metrics which the existing grid in its existing form would not be able to manage, the differences create the need for an electrical grid enhanced with digital technologies. These technologies would be designed to accept and manage electrical energy input from many different sources, with relatively different qualities, energy flows in more than one direction and the distribution of this energy in the most efficient and effective manner, such a grid would have to be intelligent and smart, thus the moniker "Smart Grid". According to the Sandia National Labs the "Smart Grid" is defined as "A modern distributed and intelligent generation and distribution network using digital technology for controls and advanced electronics for switch and protection" and is the most desired evolution of the advanced grid. Some of the features of Smart Grids as proposed by Sandia Laboratories includes but are not be limited to the following:

- Automation of power flow and energy management
- Management of the interface between the utility distributed resources and micro-grids
- Management of all power flow transitions to include effective management of multi-directional flows on the grid.
- Real time pricing and analysis for the connected communities.
- Management of the intermittency of renewable solar and wind resources.

According to the National Institution of Standards and Technology (NIST) there are ten areas that are essential to facilitate the development and growth of "SMART GRIDS" for which policies should have been created by the government and these include:

- The creation and utilization of digital technology to improve reliability, efficiency and security of the electric grid.
- The creation and utilization of technology that will dynamically optimize of all facets of grid operations and resources.
- The creation and utilization of technology that will integrate distributed renewable resources inclusive of PV Systems, CSP, Wind Systems, CHPs and Mini-Grids.

- The creation and utilization of demand response and demand-side energy-efficiency technologies

- The creation and utilization of smart metering systems and technologies that will enhance grid operations, monitor grid status and the management of the distribution grid.

- The creation and utilization of technology that will integrate smart appliances and consumer devices with the "SMART GRID".

- The creation and utilization of technologies that will integrate electricity storage and peak shaving-technologies, including plug-in electric vehicles.

- The creation and utilization of technologies that will provide consumers with timely information and controls.

- The development and utilization of interoperability standards for the SMART GRID and connected smart appliances and equipment.

- The removal of economic and other barriers to foster the adoption of smart grid technologies, practices and services.

- The creation and utilization of cyber technologies to provided enhance security for the national electric grid.

The "SMART GRID" which is expected to be the ultimate evolution in electrical grid technology, will be following in the path of earlier technologies inclusive of the advanced distribution grids and micro-grids and will be utilizing upgraded technologies that were created for those systems. The following are common elements that may be found in advanced distribution grids and micro-grids the forerunners of the "SMART GRID":

- Electronically controlled distribution systems

- Digital controls with new electronic devices providing the controls and likely performing many of the high-speed switching functions.

- Integrated electricity and communication system. The distribution system also forms the communication system which controls the distributed generation systems and an energy storage system when used. This system will also have the capacity to be dispatched to improve the system efficiency and stability, while optimizing the value of the renewable energy resource, such as a PV system.

- Integrated Energy Management System. An Energy Management System will be an integrated function when new distributed generation systems are employed and the building-Integrated electrical generation will include this energy management system to optimize the building energy generation value while providing intelligent functions such as load shedding/shifting and energy storage to provide the most value to the consumer and dispatchability for the stability of the utility. Energy management will also include intelligent and adaptive logic that manages heating, cooling and lighting needs.

- Smart end-use devices. An integrated system that utilizes communication will enable the electronic system to communicate directly with end-use devices and would automatically optimize systems operation.

- Meters as two-way energy portals. The meter and service panel for a building will be transformed into an intelligent electronic gateway that will enable electricity suppliers and customers to communicate in real time and optimize the performance and economics of the system and will interact with inverters and controllers.

- CHP distributed generation systems that enable production of both processing heat and electricity are efficient systems that can augment a central electricity generating system to improve the quality of service.

- Direct Current in Micro-Grids. The concept of power systems that generate and deliver direct current, such as PV and energy storage systems may also be revived to supply direct current loads in energy efficient micro-grid infrastructure.

Advanced Systems Architecture

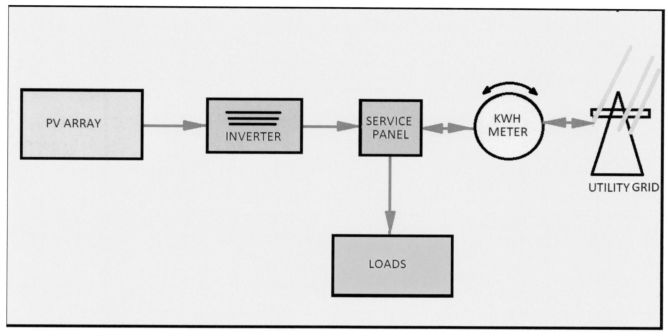

Fig.5.1 Typical Current PV System Infrastructure.

In the existing PV system configuration, net-metered system, as shown in Fig. 5.1 above, the inverter supplies power to the internal load and the grid and is metered to record net usage of electrical energy by the consumer. In this case the inverter serves three functions, DC to AC conversion, voltage and frequency monitoring and measuring and monitoring the utility power. Power from the PV array arrives at the inverter as direct current where it is converted by the power electronics of the inverter to sinusoidal alternation current, AC, as used by the grid after which the power is sent to the load and the excess injected into the grid. Due to the requirements of the IEEE 1547 the inverter must also monitor and measure the utility voltage and frequency to ensure that they are both within the design and operating specification to minimize any negative impact that fluctuating voltages and frequencies may have on the distribution system and the PV system. The inverter was also designed with power sensors and an automatic shutdown system, an anti-islanding switch that allow the inverter to disengage from the utility system if there is a utility power outage.

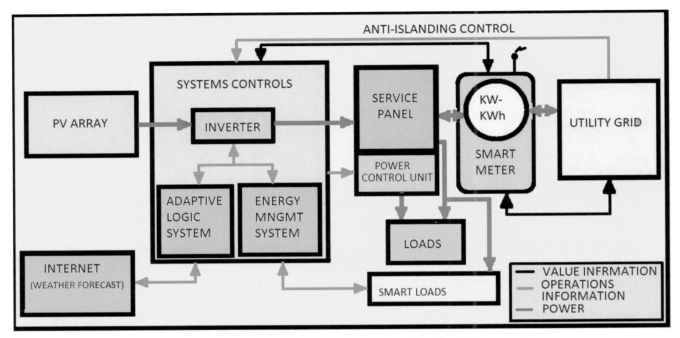

Fig. 5.2 Advanced PV System Without Energy Storage. Source: Sandia Laboratories

In the advanced PV system without energy storage, many new components are added to the system to ensure more efficient operation, more reliable power supply and greater value for the customers. These improvements are achieved by the addition of a communication system, an adoptive logic system, an energy management system, a power control unit, a smart meter, smart loads and the internet. In the new configuration the inverter has grown into a more complex and powerful unit that is designed to more effectively receive commands and information from the grid and also from the internet. In the new configuration the grid has command over the new inverter or inverter-controller and will monitor and measure voltage and frequency fluctuations and power outages and will direct the inverters, connected to the distribution system, how to react to each disturbance in the operations of the generation and distribution systems. These design changes will remove the decision to disconnect from the many inverters that may be connected to the distribution system and thereby improve the reliability of power generation and distribution systems.

The advanced PV systems can be further improved by the addition of an energy storage system as shown in figure 5.3 below, which will improve the value of the PV system and improve the reliability of the PV system operations.

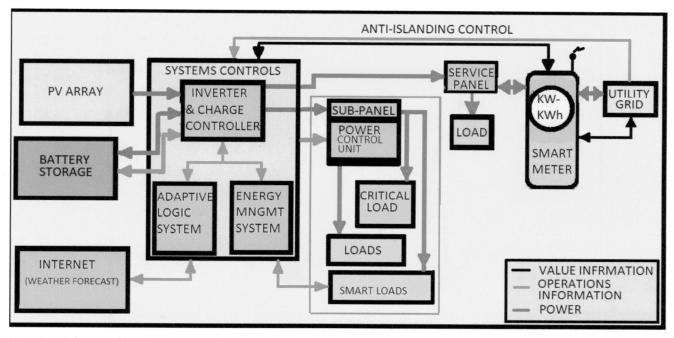

Fig. 5.3 Advanced PV System with Energy Storage. Source: Sandia Laboratories

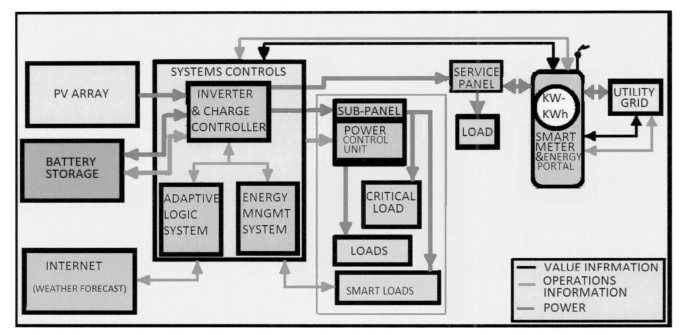

Fig. 5.4 PV System for Advanced Distribution Infrastructure. Source: Sandia Laboratories

The next evolution of the advanced PV system would involve expanding the role of the smart meter by making it into a complete energy portal system with the ability to communicate with the smart grid, the PV systems and all internal controls to ensure that solar energy will be produced and distributed in the most efficient and economic manner possible.

CHAPTER 6

The Impact of Solar Energy Production on the Environment

Solar energy has the capacity to reduce carbon emissions, which pollute the atmosphere and increase global warming, by creating alternative renewable energy systems that can replace fossil fuel energy systems and thereby help to delay the full impact of the onset of climate change upon the earth and prevent temperatures increasing above 2 C^0, however, the creation and installation of the solar energy systems of every kind also has the capacity to create their own negative impact upon planet earth, its environmental and ecological systems. This indicates that similar to all fuel types, solar energy systems while having the ability to positively impact life on earth has the capacity the capacity to also impact the earth, the natural systems that impact the earth, the animal life on the planet and the human inhabitants of the planet.

Negative Impacts of Solar Systems

Solar energy generating systems are generally divided into two major groups Photovoltaics (PV) Systems where photons of light impacts light sensitive material and causes the creation of an electrical current, the photovoltaic effect, and Concentrated Solar Power (CSP) Systems where thermal energy of the incident solar radiation is concentrated to heat water or other more efficient heat transfer fluids which can be utilized to provide high temperature high pressure steam to drive turbogenerator sets to produce large volumes of electrical energy. Both of these systems, while very different have some very basic requirements that cause both of these systems to negatively impact the environment in which they are built and operated and these basic requirements are the exclusive use of large volumes of land and the use of large volumes of water, both of which can negatively impact the surrounding environment. These are, however, not the only ways in which these solar systems can negatively impact the environment as both of these systems, while very different, have some other similarities such as the use of toxic chemicals that if released into the environment can cause significant harm. In general, solar energy systems create negative impacts in many areas inclusive of land use, water use, mining, noise pollution, visual pollution, human health, social systems, socio-economic systems, flora and fauna, and the local environment and ecology.

Land Use

Land is one of the most basic economic resource that has been and is still being utilized by all human societies since the beginning of time to create products and services to meet the most basic of humans needs inclusive of the need for food, water, shelter, clothing and the fulfilling of their most basic innate religious cravings and therefore with the great importance placed on lands any decision made by a community, a society or a country about the use of available land can create many negative impacts that could destroy that community, society and country. In general land is required for agriculture, planting and reaping of crops and animal husbandry, the reaping of natural resources inclusive of gold, silver, aluminium, iron, oil, gas, coal, undomesticated plants and animals, the reaping of water, the reaping of natural fuel wood, the provision of habitat for local flora and fauna that is important to agriculture and other areas of human life inclusive of pollinators, habitat for all the species that have made this planet their home and land for the establishment of residential, religious, institutional, industrial and commercial facilities, more so in the societies with growing populations.

This great importance of land therefore means that all decisions about the use of land must be well thought out and should only be made in consultation with all stakeholders to ensure that balance is achieved in seeking to utilize land for the production of energy and the fulfilment of the many other established demands for lands. The discussions about land use and renewable energy systems, Wind, Photovoltaic (PV) and Concentrated Solar Power (CSP), becomes critical because Wind, PV and CSP energy systems all require the use of extensive areas of land whenever there is a need to create electrical power at the utility or commercial scale, as is now the case in most modern societies that use high volumes of electrical energy. According to NREL data PV power plants require anywhere from 3.5 to 10 acres/MW, CSP power plants require the use of 2 to 10 acres/MW and onshore Wind Power Plants require even more with an average amount of 30-140 acres/MW. In addition to the extensive land requirements, most solar systems require the exclusive use of the land to ensure that the systems utilized in collecting the solar radiation is not shaded in any way, as shading will reduce the amount of energy collected, the amount of electricity produced and the overall efficiency of the solar energy systems.

The demand for high volumes of electricity in a modern society created the need for the construction and operation of many electrical generating power plants, most of which were and are still fuelled by the only recognized and highly available sources of chemical energy that could be easily converted to thermal energy, the group of fossil fuels, coal, oil and gas. Oil and its by-products were also utilized in creating mechanical power for the many elements of the very important transportation systems utilized all over the world and transportation systems are responsible for a very large portion of the carbon dioxide in the atmosphere. Unfortunately, this group of fuels had a major drawback, the emission of global warming substances, the effects of which were not readily recognized and acknowledged until the emission started creating global problems within our atmosphere and consequently our climate. While human caused climate change is only now being recognized, the advent of human caused climate change was commenced very early in human existence as the many deserts across the world and decimated forests and species of plants and animals will testify to, but this problem was exacerbated in the last two hundred years by the industrial revolution that created many new methods of production which were and still are dependent upon electrification created by "cheap energy", urbanization that created a demand for many new electrical household goods and tools and the overdependence of modern economies of developed and developing countries on these "cheap fuels", fossil fuels. These "cheap fuels" have been responsible for the emission of humongous volumes of carbon and other global warming substances into the atmosphere that has consequently led to the onset of a global climate change with the possibilities of global cataclysms. These possible outcomes have forged the need for an urgent and great shift away from the use of fossil fuels to fuels that will not pollute the atmosphere and consequently not impact the ability of the envelope containing the earth, the atmosphere and all of the associated environmental climate systems, to self-regulate.

The acceptance by modern societies of the need to shift away from the use of fossil fuels towards clean renewable sources of energy to meet global electrical energy demands means that modern societies must also accept the need to make modifications to the ways that they have done things since the start of the industrial revolution, things that created social and economic upheavals that changed the very nature of how societies were organized and operated during that era. Prior to the industrial revolution agriculture was the economic main stay of most societies and a very large portion of the population was employed at one agricultural activity or another, however, after the revolution this changed significantly with a large portion of the population getting "jobs" in factories which led to people becoming less physically attached to the land while at the same time remaining sentimentally attached to it. In that period prior to the revolution most people also lived in rural communities close to their jobs on the farm and in many other cases, they lived directly on the farm because the farms were owned by the families that operated them. This reality was also disrupted with the creation of new "work" by the industrial revolution and people started moving away from the farms to live closer to their new "work", usually in close proximity to the facility in which the new "work" took place. Over time the old societal norms continued changing to meet the changing realities of most modern societies, however, most societies retained their emotional and sentimental attachment to land, a sentiment that sometimes blocks the need of making required changes that are designed to make the society safer. One such change that must now be made is the transition away from fossil fuels to renewable sources of energy to produce the electricity that all modern societies demand, however, this transition requires the utilization of vast quantities of land that many are still emotionally attached to, as land is still seen as that most precious resource that all humans need in order for them to be able to feed themselves, even in a time when food production is much more efficient and much less land is required to feed a growing human population, today much more food is being produced than at any other time in human history.

With the growing importance of renewable energy in the age of climate change, renewable energy systems' demand for large volumes of land has begun to compete with the traditional uses of land in many areas of the world, including in the developed and developing world, and this demand has led to agricultural land being converted for use in creating greatly needed renewable sources of energy. In the US for example, agricultural lands are now being converted to create renewable energy farms and this conversion is giving farmers hard hit by low crop prices and other economic pressures related to agriculture, a new avenue to make money. According to Hall et al (2022) there has been an exponential growth in the number of food farms that have been converted to renewable energy and solar energy farms, since 2009. The number of food farms converted to renewable energy farms are as follows 2009 – 9, 509 farms, 2012 – 57,299 farms and 2017 – 133,176 farms. Of these, the majority are conversions to solar energy farms, 2009 -7,968, 2012 – 36,331and 2017 – 90,142. The number of food farms that will be converted to solar energy farms are expected to increase even more, as IRENA projects that the use of renewable energy is projected to grow from 16% to 66% by 2050 and while most people would prefer that only marginal lands be used for these renewable energy projects, the developers of these renewable energy systems prefer good, flat, farmlands which makes it much easier to construct their large projects.

While these conversions have all been done for the economic benefits that they provide to the owners of the lands, the provision of cost effective clean energy and filling the great need to utilize more renewable energy systems, many sections of US society including many local authorities, many NGOs and community groups that are very concerned about the preservation of farm lands, are in disagreement with this land conversion because as stated by Owley and Morris (2019) "Loss of agricultural land can have significant effects on the economy, the environment and the social fabric of a community". This great fear of the loss of land with food producing capacity is an age-old fear that is an innate part of all human beings, a survival fear that will not easily go away and in most cases, the authorities in most locations are in agreement with programs that promote the increased use of solar energy and they promote these programs to ensure a reduction in dependence upon fossil fuels in order to reduce and or eliminate the level of pollutants in the atmosphere which are responsible for global warming and climate changes.

The use of land also significantly impacts an area of life that people in developed economy tend to place very little economic value on, this area is the religious or spiritual value of land which has cause several conflicts between Native Americans and the economic owners of oil pipelines, oil wells and several other activities that do not show respect to their sacred lands. According to Barclay and Steele (2021) "The threat to sacred sites and cultural resources continue today in the form of spoilation from development, as well as in the significant barriers Indigenous people face in accessing and preserving these sites and resources." Native indigenous people face this sort of problem all around the world as the government of their countries seek to utilize their sacred lands for the economic and environmental benefits of the "nation" without seeking the input or permission of the native indigenous people. In India for example indigenous people are having a problem with energy companies placing solar power plants on their sacred lands and according to Gawande and Chaudhry (2019) a 235 MW Solar Energy Power Plant was built on sacred lands which the people of the community had reserved for a local goddess. This resulted in conflict between the project developers and the local community even though the developers had been given the legal right to develop the project the by the relevant local authority and Ministry responsible for such matters in the Indian Government. The community people were disappointed and angry and said "It is not like we could not have used the land if we wanted, parts of it could have been farmed but it is sacred land and no one else can use it", this statement formed the basis for conflict and resistance which probably could have been avoided with proper consultations. The community's anger was compounded when their holy trees were also destroyed to make way for the energy project.

While the use of farm lands to produce renewable energy is an essential ingredient in the fight against climate change, the use of the land in this manner has several drawbacks inclusive of the temporary loss of the land for agricultural purposes and the loss of habitat for many species of animals and plants, which will greatly impact the biodiversity and the sensitive ecosystems of the areas in which solar power plants are built. There is also the risk that land being used for the creation of solar energy power plants may become contaminated by the many toxic materials that are used to construct PV solar cells, heat transfer fluids for use in CSP systems, battery parts, and harmful toxic heavy metals that may be used in wind turbines, thus causing the permanent loss of these lands for agricultural and other community purposes. The loss of agricultural lands caused by pollution from toxic heavy metals is already a reality in many parts of the world that have a heavy inflow of electronic waste into their society.

Water Use

Water is one of the most essential elements for human existence and survival and the loss of any potable water source, for any reason or any measure, is considered a major problem that must be resolved at the earliest possible time. The use of large volumes of water has been an integral part of the processes utilized to create electrical energy in most traditional electrical power generating plants, where water is used for the generation of high temperature and high temperature steam to drive turbogenerator sets and for the cooling of heat exchangers such as the condensers for condensation of steam to create the necessary low pressure at the outlet side of the steam turbine. While solar systems, PV and CSP, do not require combustion of any fuels they both require large volumes of water in their daily operations and CSP systems may require even a little more water than the traditional power generating plants as in addition to water utilized in steam generation, it also utilizes water for cleaning mirrors.

The CSP power plants utilize water for the same reasons as traditional power plants, cooling of heat exchangers for the condensation of steam to create the necessary low pressure on the outlet side of the steam turbine, for absorbing and transferring heat generated in the solar concentrators to the turbogenerator sets where the steam created expands to move the electrical generator's rotor coils in the generator's stator across a rotating magnetic field established by the stator coils. Water is also used extensively in the maintenance of the CSP power plant and according to the Solar Energies Industries Association (SEIA) the amount of water used varies between 500-850 gallons/ MWh of power generated within the plant, similar to traditional power plants and approximately 20 gallons/MWh to keep the mirrors and or reflective surfaces clean. These amounts do not include the amount of water that would have been utilized in the manufacture, construction and installation of the components that make up a CSP plant.

The PV power plants do not use water in the process of generating electrical power, however, large volumes are used on a daily basis to clean accumulated dust from the solar collectors to ensure the optimum collection of incident solar radiation and According to the SEIA the amount of water utilized in that manner on a daily basis is approximately 20 gallons/MWh. The amount of water utilized in PV and CSP plant may be relatively small compared to the use of water in other industries, however, in locations where water is scarce even this small amount of water could create significant hardships on the inhabitants of the community in which the plant is located. This "necessary" use of water in PV and CSP power plants may also create unnecessary conflicts between the people of the community and the owners or operators of the solar power plant as in regions where well water is used this usage could even lead to significant damage to the aquifers and surrounding lands. Water withdrawn from local rivers, streams, lakes and underground aquifers may also significantly impact the ecological system of these water bodies, the surrounding dependent systems and consequently impact the food supply of a community or region as many communities subsist on aquatic plants and animals. The reality is that rivers and other surface water bodies need to be maintained at minimum flow levels to support the life of all the species that depend on it inclusive of aquatic life like fishes, crabs, shrimps, oysters, salamanders, turtles and tortoises, and the aquatic plants that support both human and aquatic animal life. These surface water bodies also support many species of avian life that find habitat directly on the banks of these water bodies or in the forests and savannahs that occupy their banks or on islands that may form in the middle of large rivers and lakes. Therefore, it can be seen that excessive water withdrawal could lead to the decimation of many species and negatively impact the lives of the human communities that grew up along river banks or around the shores of lakes or on islands. Some of these negative impacts could include the destruction of local economies dependent upon the body of water, the loss of transportation systems as the waters become unnavigable, the loss of a meaningful livelihood for the local population and the eventual loss of the community itself, as the local population would be forced to move away to stay alive.

Hazardous Material

Hazardous material used in the Solar energy industry fall into four categories, solvents and cleaning chemical, toxic materials used in the manufacturing of PV cells, toxic heat transfer fluids and heavy metals used in the batteries that are so essential for solar systems.

Cleaning Chemicals. The manufacture of PV cells requires the use of several hazardous materials inclusive of hydrochloric acid, sulfuric acid, nitric acid, hydrogen fluoride, 1,1,1- trichloroethane and acetone, all of which if mishandled can significantly pollute the environment and cause physical harm to the people working with and around them. These chemicals are generally utilized during the process of manufacture of the PV cell and are used to purify and clean the surfaces of the semiconductor, PV cells, and the volume of these material utilized at any one time will be a function of the type of cells, the amount of cleaning and purification and the size of the wafers that are being manufactured. All of these chemicals can be quite harmful to both humans and the environment as follows:

- Hydrochloric Acid. Hydrochloric acid has found many uses in the industrial, commercial and residential sectors, however, it is hazardous, toxic and corrosive and long-term occupational exposure to it can lead to several illnesses in humans inclusive of gastritis, chronic bronchitis, dermatitis and photosensitization.

- Nitric Acid. Nitric Acid has many uses in many industrial and commercial entities; however, it is highly corrosive and can have significant negative impacts on people with long-term occupational exposure which can lead to several diseases in humans inclusive of delayed pulmonary edema, pneumonitis, bronchitis and dental erosion. It also impacts the environment in several ways, it is one of the main chemicals that is used in the formulation of fertilizers and it also reacts to form nitrogen oxides that are injurious to the natural environment.

- Hydrogen Fluoride. Hydrogen Fluoride or Hydrofluoric acid is extremely corrosive, toxic by inhalation, highly toxic by ingestion and highly toxic by skin absorption, highly destructive to human flesh and bones as it can quickly penetrate skin, flesh and bones, will damage lung tissue, cause swelling and fluid accumulation in the lungs (pulmonary edema)

- 1,1,1,-trichloroethane. This is a volatile synthetic chemical that has found wide use in industry and on the domestic front where it is used as a solvent or in the making of things like glue. It is harmful to the ozone layer, its production is banned in some areas of the world, and it can also be harmful to the health of humans if inhaled in high concentrations where it can cause dizziness, unconsciousness, lowering of the blood pressure and the stopping of the heart. In addition, it will also damage the breathing passage, damage the nervous system and negatively impact the liver of humans.

- Acetone. Acetone is found naturally is trees, fruits and even in humans, but is manufactured in high volumes for industrial, commercial and residential uses and it is a solvent that is volatile and very flammable, and it can greatly impact the health of both humans and the environment in a very negative way. In humans the inhalations of different levels of acetone can cause significant health distress such as headaches, nausea, racing pulse, confusion, changes to the size and number of blood cells, unconsciousness and coma. In the environment acetone in the atmosphere can be broken down by sunlight to cause the formation of ground level ozone which is a criteria pollutant.

Solar Cell Material. Some of the material currently used in the construction of solar cells are toxic and includes material such as silicon, indium, gallium, germanium, arsenic, copper and cadmium and some of them are toxic and have negative impacts on both humans and the environment.

Silica/Silicon. The major component in 90 percent of all PV cells is silicon which is derived from natural silica, sand or quartz, and both sources can negatively impact the health of the people with prolonged exposure to the fine breathable particles during the process of manufacture. According to the US CDC silica has been shown to negatively impact the health of humans through inhalation of fine particles of silica over a prolonged period of time and this can lead to several diseases inclusive of silicosis a fibrotic lung disease, chronic obstruction pulmonary disease (COPD), lung cancer, renal toxicity, increased risk of tuberculosis and autoimmune diseases. The reduced silica, silicon likewise can have negative impacts on human health when humans have prolonged exposure to the fine breathable particles and will also have the following health impacts, chronic respiratory effects, irritation of the skin and eyes, irritation of the lung and mucus membrane.

Rare Metals. The other type PV cells, thin film PV, that is currently used only in about ten percent of PV cells, are usually made from exotic rare metals inclusive of germanium, indium, cadmium, gallium and compounds of these rare metals inclusive of gallium arsenide, copper-indium-gallium-di-selenide/sulfide, cadmium telluride and amorphous-silicon/amorphous-silicon/ amorphous- silicon-germanium, some of which are quite toxic and have the capacity to be injurious to health of both humans and the environment. Some of these metals will have little to no effect upon humans, while other will, these are as follows:

- Gallium. Ggallium will cause dermatitis and gastrointestinal problems and gallium compounds such gallium arsenide are toxic.

- Germanium. Germanium is relatively benign but two of its compounds, germanium hydride is toxic and germanium tetrachloride can hydrolyse in air to form a very corrosive Hydrofluoric acid.

- Cadmium. Cadmium is considered by the USEPA to be a probable human carcinogen.

- Arsenic is a naturally occurring substance and according to the World Health Organization (WHO) it is highly toxic and if consumed in significant quantities may lead to instantaneous death and if consumed in water or food over the long-term may lead to several significant diseases inclusive of cancer, skin lesions, cardiovascular diseases and diabetes.

Heat Transfer Fluids. Some CSP systems utilize heat transfer fluids inclusive of glycol-based compounds, glycerol-based compounds, synthetic oils, molten salts and molten sodium and some of these compounds are toxic and very corrosive and they are consequently potentially harmful to human health and the health of the environment. The Safety Data Sheets, SDS, of two of the more popular synthetic oils heat transfer fluids, Therminol VP1 and Dowtherm A clearly indicate the negative impacts they have on humans, fishes and aquatic life, with humans they are harmful if inhaled or swallowed and they are deadly to fishes and aquatic life. The molten salts used in CSP systems are usually a mixture of 60 % sodium nitrate,40% potassium nitrate and or calcium nitrate all of which have been shown on their SDS to be hazardous, toxic and in one case is a probable human carcinogenic and will also negatively impact the wider environment.

Battery. Many solar systems utilize batteries which are usually made with lead, cadmium, nickel and lithium, all heavy metals which are quite toxic and harmful to both humans and the environmental systems and a significant environmental problem arises when these batteries are to be replaced and disposed of. According to the National Institute of Health (NIH) they have the following negative impacts:

- Lead. Lead is toxic and exposure to it can negatively impacts the health of humans in several ways inclusive of affecting the mental development of children, causing problems with pregnancies, impacts human blood pressure and hypertension, affects the functioning of the human kidneys and is also a probable human carcinogen.

- Cadmium. Cadmium is toxic and is also a known carcinogen that may be responsible for various cancers inclusive of breast, lung, prostate, nasopharynx, pancreas and kidney cancers.

- Nickel. Exposure to Nickel can negatively impacts the health of humans in several ways inclusive of allergies, cardiovascular and kidney diseases, lung fibrosis, lung and nasal cancers.

- Lithium. Exposure to Lithium can negatively impact humans in several ways inclusive of eye, skin, nose, throat and lung irritation, build-up of fluid in the lungs (pulmonary edema), loss of appetite, nausea, vomiting, seizures, coma, impact the thyroid gland, kidney and heart functions. Lithium is also an explosive hazard and is quite toxic.

Local Climate

Renewable energy has been widely acknowledged as one of the major solutions for reducing the impact that global warming substances, carbon dioxide and other products of the combustion of fossil fuels which are emitted into the atmosphere and other industrial activities, were having on the global climate and while the utilization of renewable energies have and will continue to have a positive impact on the reduction of global warming substances in the atmosphere, some of these systems, in particular solar power plants, may also add to the heat in local communities in which they are located. Each solar energy plant, PV and CSP, would consists of thousands of solar collectors and mirrors which were designed to absorb more energy in a particular location than would naturally be the case. However, according to many sources the impact on soil and the ecology of the location is not significant and is not lasting.

Solar Equipment End of Life

The projected life span of most solar systems is in the range of 20-30 years and at the end of this period the panels and other systems must be replaced as the capacity of the collecting cells and mirrors would have deteriorated over time and this deterioration reduces their ability to continue producing electricity in an efficient and cost-effective manner. Secondly, during that 20-30 years period of time many newer generations of more efficient and cost-effective PV systems would have been created making the existing 20-30 years old systems obsolete and unprofitable to use in solar power plants. This scenario creates the need for several solutions to manage the safe long-term storage of material that cannot be recycled and the recycling of everything that can be reused or repurposed. The location and facility to store material waste from solar power plants that cannot be recycled could contain material that has the capacity to contaminate and negatively impact the location and surrounding environment, especially land, surface water and ground water resources.

Social Impact

Farming has been a way of life in communities around the world from time immemorial, beginning when humans started giving up their hunter-gather lifestyle, many millennia ago, for a more stable existence that began with the production of their own food by their own hands, an activity that gave all men so engaged a sense of confidence and independence that has lasted throughout time. The sense of satisfaction achieved by these early farmers and their sense of independence told these early humans that living and surviving by the sweat of their brow was the most satisfying thing in the world as their hard work, in good and bad times, provided the most essential things in life and the physical work associated with producing their own food also kept them physically healthy as they slept well and was able to enjoy the beauties of nature when they did not have to work.

In their development of agriculture and the technologies associated with it these early humans learnt how to fight with the elements, learnt the proper times to plant and reap many different crops from which they developed calendars and they domesticated many species of plants and animals, and from all of these activities they developed a bank of knowledge that they recorded over time and they had passed on the knowledge gained in each generation to the next generation. They passed on the knowledge gained about plants and animals, seasons and time, rain and water systems and they also passed on their sense of achievement and independence, that had kept their families, villages and towns alive through many millennia. This sharing of information among families and communities and the time spent together, formal and informal, sharing the accumulated information helped to build very strong familial and community bonds that lasted for generations. All the information, passion and sense of community passed on to future farmers would be of very little use, however, if the future farmer had no land on which to practice his craft, build a family, build a community and continue the many traditions started by his forefathers. This future farmer would be like the proverbial fish out of water as land was and is the most important ingredient in building and sustaining quality human life on planet earth.

Over time most early humans had become farmers and the population of hunter-gatherers dwindled, but never completely disappeared, and the majority of the human population became attached to the land and not just any piece of land, but the specific piece of land that their families had gotten attached to through one or the other social transactions from ancient times inclusive of bonds of slavery and serfdom, family purchases, family acquisition by other means, endowments by kings or queens or stolen from the monarchy or government. Historically not all farmers remained stable as some were very mobile as they would move from place to place, never securing and holding tenure to the land while still practicing their farming and even while mobile the attachment to the land was still very strong continuing the practice that all ancient families, communities and villages were all built up on the land and around farming, creating very strong bonds that lasted through many millennia into modern times.

This attachment to the land went very deep into the human psyche and may have even attached itself to our very DNA and losing your farming land would become the end of a family, communities and villages as has happened on many occasions over time with invasion by more aggressive tribes of humans. This loss of land quite often separated families, communities and destroyed the social bonds developed over time, bonds that were very important to the well-being of all families and communities as all depended on each other for familial, emotional, psychological, religious, communal, work and even financial support to overcome and thrive every planting and reaping season and even during harsh times when they could do no work. These truths are acknowledged by Owley and Morris (2019) "Loss of agricultural land can have significant effects on the economy, the environment and the social fabric of a community".

Socio-Economic Impacts

As acknowledged above, throughout time some very strong bonds were developed between the land, the community, their social and economic activities and the severing of these bonds usually had many negative impacts, social and economic, upon the lives of the people involved. The coming of the industrial age created a new reality that allowed the breaking of these bonds with the land and for a significant percentage of the human population this new reality allowed many to leave the land and still have viable economic activities that allowed them to maintain their social bonds. The industrial revolution, however, could not convert the whole human population to factory workers and office staff as the world still needed food that the factories could not produce, food that only the remaining farmers could produce, farmers with a greater attachment to the land than those that had left. Unfortunately, over time many of these dedicated farmers were forced off their lands by the economic activities of those that had lost their attachment to the land and this created even more devastation to farming families as even more communities were destroyed and many farmers were forced into poverty with the loss of their land and their ability to perform their economic activity. During the 1929 economic depression that affected the world farmers suffered and according to newspaper reports and other data farmers all around the globe suffered from the following:

- People could not afford to purchase many basic food items and in response to this food prices fell leading to greater problems for the farmer.

- The falling food prices force many farmers into bankruptcy as they could not meet their financial obligations to banks and other institutions to whom they owed moneys such as mortgages and crop loans.

- In the US the tax man also took the land of many farmers as they could not even afford to pay their taxes.

- Tennant and share-cropping farmers suffered even more as they had no secure tenure on the land that they farmed and in order to survive they had to abandon farming and move into the cities to find jobs in factories and similar places.

- The families of all farming communities were hard hit and many moved out of farming forever, breaking family traditions and weakening the strong social bonds that had been built up over many millennia.

- Many farmers were also forced to do the unthinkable, destroy their crops and animals while millions were hungry, under the instruction of the government in order to reduce the availability of farm goods on the market and thereby force up the price of these goods.

- Family farms were replaced by large industrialized farming businesses all staffed with paid labor just like in the factories.

The effects of the farmer's loss of agricultural land to facilitate the growth of the renewable energy industries is in some ways similar to the losses that occur during the Great Depression, except that in the renewable cases the land-owning farmers are profiting, while in many places the farmers without land tenure are again losing everything. Many people in many countries around the world without secure land tenure will never benefit in any way from the new green industrial revolution as the governments of their countries have confiscated their traditional farms and holy lands to hand over to the energy companies that will be building and operating the renewable energy plants. The confiscation of the community's lands by the government, due to the governments new energy policies, caught many villagers by surprise as many only learnt at those critical points that they had no secure tenure on the lands that they may have lived and farmed on for centuries. They learnt at these critical times that because they had no certificate of title, a condition which was never required in their traditional land ownership practised by their fore parents and ancestors, they did not have any hold on the land in a modern economy. These situations occur most often and is most visible in the rural communities of the world where the people may be illiterate and lack proper representation within the halls of politics.

These populations are also further abused by the political and social systems when they have no knowledge of modern construction technology, have no knowledge of solar energy systems or any modern technology which automatically makes them unemployable by the renewable energy project owners and contractors. This absence of technically trained people within the selected communities would automatically lead to the employment of people from outside of the rural communities, most likely coming from the nearest cities or universities and this inflow of new people into these communities would require further changes and disturbances that the people of the community would have no control over and may not even benefit from in any way. The required changes to community may include the construction of new infrastructure residences, schools, health centers and shopping centers, all of which could bring significant physical and social changes to the indigenous or existing people of the community, changes that the existing community members may never benefit from directly or indirectly in the immediate situation.

Noise Pollution

Most solar PV systems will never make enough noise during operations to disturb anyone, however, significant amount of noise will be made during the manufacturing process and onsite during the construction and maintenance of these systems. The same thing however, cannot be said about CSP systems as these systems will utilize the heat that they generated to make high pressure steam which require the application of safety devices similar to other steam power plants. These safety devices, such as pressure relief valves, will at some point during operations allow a full flow of steam for very short intervals and any release from a high-pressure steam system usually produce very loud noises. These safety devices also require intermittent exercising to ensure that they are in good operating conditions and these activities are done on annually, semi-annuals or as is required by the jurisdiction in which they are located. The construction of these plants will also generate significant levels of noise during the manufacturing process, onsite during construction and during maintenance activities.

Visual Pollution

Nature produces everything in forms that are pleasant to the eyes and other senses, no matter how bare or how stark, they appeal to the natural senses of all human beings, however, when the natural landforms, plants and the other elements are transformed or covered with concrete, wood and other building material the space can lose its natural aesthetic appeal unless special effort is made to ensure that the built structure does not impede the natural beauty of the site or have a minimal impact. A thousand acres of corn is quite appealing to look at from above when flying, or from the side when driving beside them on the road, how however, the same 1000 acres covered with solar panels, mirrors and collectors or wind turbines just does not impact humans in the same way as corn does and most humans would rather not see them especially in the communities in which they spend most of their time. These renewable energy systems are aesthetically very unappealing to 90 percent, at least, of most humans, who find them to be an eye sore and based on this very strong dislike most will protest the location of renewable systems anywhere near their home communities. Many renewable energy projects have had their permits to build rejected or had their approval revoked after protest by communities, however, this rejection by most communities does not mean that the communities do not desire to have clean energy to use, it just means that they do not want the means to produce the clean energy in close proximity to their communities and homes, they should be somewhere out of sight.

Comparative Negative Impacts

All power plants negatively impact the environments in which they are built, especially in areas such as land use, water use, pollution, social and socio-economic ways. A comparison is shown in the table below.

Energy Type	Land Use Acres/MW	Water Use Gallons/MWh	CO_2e Pollution Tonnes/GWh	Environmental Pollution	End of Life Pollution
Coal	12.21	1,005	888	Overwhelming	Significant
Natural Gas	12.41	255	499	Significant	Low
Nuclear	12.71	1,101	29	Low	Significant
Solar PV	10	70	85	Low	Medium
Solar CSP	10	906	80	Low	Medium
Wind	70	88	26	Low	Low
Hydropower	315.22	18,000	26	Significant	Low
Geothermal	6.0	15	38	Low	Low

Table 6.1 The Impact of Electrical Generating Power Plants on Land and the Environment.

Positive Impacts of Renewable Energy Systems

The current climate crisis has its roots in anthropogenic activities, activities that stem from the industrial revolution, other large scale human activities and the creation of new industries and machinery that were heavily reliant upon electrical energy that is produced from different sources of energy inclusive of wood, coal, oil and later natural gas, fuels which all polluted the atmosphere with carbon dioxide. The atmosphere typically consists of several gases inclusive of nitrogen 79%, oxygen 20%, carbon dioxide 0.035%, methane, particulate matter, sulphur dioxide, ozone, water vapor, and several other gases in minute amounts. Carbon dioxide, water vapor and methane along with a few other gases, are essential global warming substances which worked together to create and maintain a liveable average surface temperature of 59^0 F on Earth which allowed the evolution of an environment that facilitated the development and growth of plant, animal and human life on the surface of the Earth and without these gases the temperature of the earth would be an average of 0^0 F which could not sustain plant and animal life.

Information from the National Ocean and Atmospheric Administration (NOAA) indicates that the natural level of carbon dioxide in the pre-industrial revolution atmosphere was approximately 280 parts per million and that from the beginning of the industrial revolution to present times this level has risen to approximately 420 parts per million, 140 parts per million above the preindustrial age levels. According to the USEPA the excess carbon dioxide in the atmosphere is due to anthropogenic activities inclusive of agriculture 11%, commercial & residential 13% activities, generation of electricity 25%, industrial activities 24% and transportation 27%, with the carbon dioxide being produced by a few processes inclusive of combustion of carbon rich fossil fuels 87%, clearing forests and land use changes 9% and industrial processes 4%. The natural capacity of the carbon dioxide to absorb and reradiate the heat of the sun ensures that the excess carbon dioxide levels in the atmosphere will lead to an increased amount of the sun's heat being absorbed and reradiated towards the earth and the atmosphere and this will lead to the heating up of the earth and surrounding atmosphere beyond its natural levels. The result of this increased heating, which occurs at a global scale, will create many changes in the behaviour of the earth's climate and the impacts will see many changes inclusive of the warming and expansion of sea water which will lead to the raising of sea water levels and the melting of permanent ice caps, sea ice, permafrost, glaciers and other formerly permanent bodies of ice. The resultant projected increases in sea temperatures and levels will negatively impact aquatic flora and fauna, flooding of coastal regions inclusive of coastal cities and negatively impact human life, as these coastal cities currently house a very large percentage of the human population of most countries. The rise in sea water temperatures and levels will also impact land-based flora and fauna, especially farms and forests on the coastal regions, and thereby impact food production and the availability of food to feed the growing human population.

According to the USEIA, electrical power generation plants utilizing fossil fuels were responsible for 62% of electrical energy production, but for 99% of the carbon dioxide emissions related to the production of electricity. The demand for electrical energy is growing in all regions of the world and these demands are mostly being met by the combustion of carbon rich fossil fuels and according to the USEIA, in 2021 fuels used in the US to produce electricity were Coal 22%, natural gas 38%, oil 1%, nuclear 19 %, Hydro 6.3 %, wind 9.2%, Biomass, Solar 2.8%, and geothermal 0.4%. According to Bloomberg 2021 data, in China electrical energy was produced by fossil fuels 71.13 %, nuclear 5%, wind 6.98%, Solar 2.25% and hydro 14.59% and according to the Ministry of

Power, India, electrical energy was produced by fossil fuels 58.6 %, hydro 11.6%, wind 10.1%, solar 14.1%, small hydro 1.2 % and other 2.5%. This data clearly shows that fossil fuels inclusive of coal, oil and natural gas are still being used in large quantities and if new carbon free sources are not found and utilized the expected growth in the demand for electricity over the next 50 years will continue to be met by these carbon rich fossil fuels and the problems associated with them will only exacerbate the climate crisis.

There are currently several established carbon free sources of energy inclusive of solar, PV and CSP, Wind, onshore and offshore, Biomass, solid and liquid, Hydro, river and ocean based, geothermal, Hydrogen fuel cells and Nuclear and the ones with the least carbon dioxide equivalent impact are as shown in the table above are Solar, Wind, Hydro, geothermal and nuclear. When all of the impacts and the long-term opportunities are considered, solar energy has one of the best outlooks relative to nuclear, hydro and geothermal sources of energy for several reasons inclusive of, the possible long-term effects of the waste from nuclear power plants, the environmental, ecological and hydrological damage that hydropower plants have wrought and the limited nature of geothermal sources around the globe. Solar energy is available everywhere, though not in equal amounts everywhere as some locations will have more sunlight for longer periods of time than others during certain times of the year and it is more available in the regions around the equator and areas in close proximity to this zone. The availability of a national electrical grid makes it possible for the sun rich regions of a country to supply clean electrical energy to the other regions that are not so well served, especially the regions with the largest populations where the need for clean electrical energy is in greatest demand.

According to the USEIA the global installed electrical energy capacity in 2020 was 7742 million kW (7742 GW) with a total capacity of 40,691, 952GWh operating at 60% of capacity and the carbon dioxide equivalent emissions required to produce the required electrical energy in 2019 was 35.55 billion metric tonnes (35.55 Gt). The energy sources currently being used to create this 7742 GW electrical energy are Fossil Fuels 4415 GW, Nuclear 394 GW, Wind 736 GW, Solar 716 GW, Hydro 1162 GW, Biomass 136 GW, Hydro Pumped Storage 168 GW and Geothermal 14 GW, with the Fossil Fuels producing all of the negative emissions. These numbers highlight the great opportunities that continues to exist for the reduction or elimination of carbon dioxide emissions with the replacement of fossil fuels with carbon-free sources of energy over time. In this vane many countries have created projections to increase the levels of solar energy that they will utilize from the current levels of 5%, to levels as high as 50% by 2050. IRENA also projects that by 2050 Solar PV systems should be supplying 42.59% of global electrical energy demands and wind should be supplying 30.22% which when combined would eliminate a very large percentage of all global carbon dioxide emissions produced from electrical power generation.

This very large projection in the reduction in carbon dioxide emissions by 2050 would create a very positive impact on global climate conditions and would be an indicator that the positives from solar energy systems outweighs the many negatives identified earlier inclusive of land and water use and to achieve the needed growth in utilizing solar energy on the scale projected will require two things, a paradigm shift in how humans think about the use of land and water and a shift in the current designs of PV solar systems that could reduce the amount of land required. There also needs to be a shift to what is now being described by many in the field as agri-voltaics, a system that is designed to facilitate the utilization of land for PV systems and agriculture at the same time, as such a system could encourage more farmers to get involved in solar energy production as a second source of income.

CHAPTER 7

Solar Geometry

Solar geometry speaks to the many angles at which sunlight impinges upon the surface of the Earth, solar energy collecting surfaces and the intensity of the solar energy at each point of impingement and consequently the angles at which all solar radiation gathering systems must be placed to facilitate maximum utilization of this great source of energy.

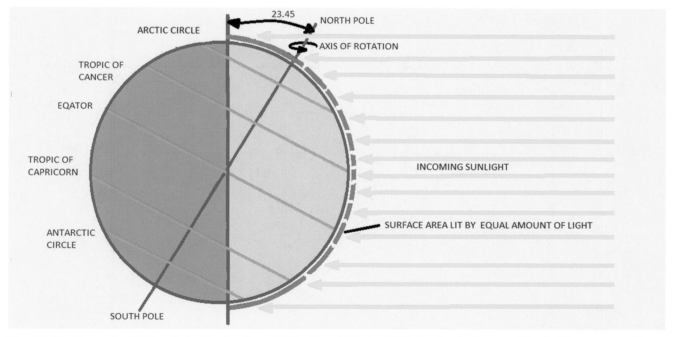

Fig. 7.1 The Intensity Of Sunlight On Surface Of The Earth Varies With Latitude. Source: Earth Observatory, Nasa.

The total solar radiance is the maximum power delivered to a surface assuming that the path of the incident incoming sunlight and heat is perpendicular to that surface and the spherical shape of the earth ensures that only places that are at or near to the equator receives sunlight at this optimum angle, this perpendicular angle. Consequently, the locations at or close to regions around the equator are the places that receive maximum solar radiation as most other locations only receives incident sunlight at an angle less than the perpendicular and solar irradiance decreases with the decreasing angle of incidence. The maximum impact in this region would usually occur during the periods of the year when the angle of declination is zero, March 21 and September 21 as the earth moves through its orbit, as shown in Fig 7.3 below. This fundamental fact impacts the designs of all systems that are designed to collect and utilize incident solar energy and each location on the face of the earth must be treated based on their location relative to the equator. This very important angle of incidence also changes as the sun moves across the sky and consequently impacts the amount of solar energy that is collected by solar panels and other solar systems throughout a standard 8-12 hours of sunlight each day.

Movement of the Sun Across the Sky

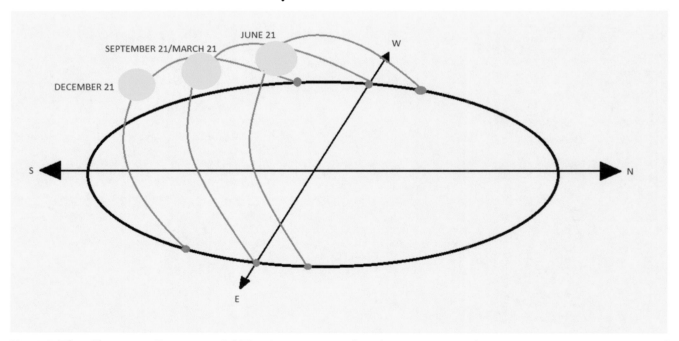

Fig 7.2 The Changing Positions Of The Sun Across The Sky. Source: Kalogirou, Cyprus University Of Technology.

The impact of solar energy on the surface of the earth is also significantly affected by the movement of the sun across the sky as shown in FIG 7.2 above, which shows the Sun in three positions based on the time of the year and the particular point that the earth is in its orbit around the sun. These dates are December 21 the Winter Solstice, June 21 the Summer Solstice, March 21 the Spring Equinox and September 21 the Fall Equinox, all very important times on the earth's calendar.

Solstices

There are two solstices each year, June 21 the Summer Solstice and December 21 the Winter Solstice and a solstice is defined as those times of the year when sun reaches it maximum and minimum declination, relative to the earth and this results in the longest day of the year on June 21 at the maximum declination and the shortest day of the year on December 21 at the minimum declination. The solstices also mark the time when the Sun reaches the highest point in the sky, June 21, and the lowest points in the sky, December 21 as shown on FIG 7.2 above. These days also mark the beginning of Summer in the northern hemisphere, and winter in the Southern Hemisphere, June 21, and the beginning of winter in the norther hemisphere and summer in the southern hemisphere, December 21.

During the Summer Solstice, June 21, the Sun shines down directly on the Tropic of Cancer in the northern hemisphere, where its rays make an angle of 23.45^{0} with the Earth's equatorial plane and this angle is known as the declination angle and is defined as the angle formed between the rays of the sun, extended to the center of the earth and the earth's equatorial plane. During the Summer Solstice the Sun is at its highest point in the sky and is above the horizon in the northern hemisphere for the longest period of time, while at the same time it remains below the horizon in the Antarctic Circle in the Southern Hemisphere. During the Winter Solstice, December 21, the Sun Shines most directly on the Tropic of Capricorn, in the Southern Hemisphere, and areas of the Northern Hemisphere experiences significant periods without direct sunlight as the sun remains below the horizon in the Arctic Circle, as shown in FIG. 7. 3 below.

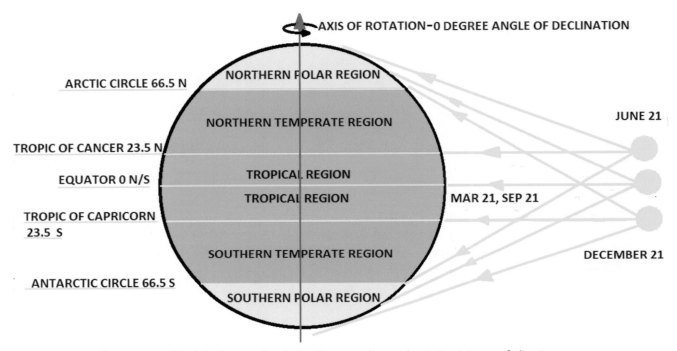

Fig 7.3 Typical Diagram Highlighting the Solar Impact from the 3 Positions of the Sun

Equinoxes

There are two days, March 21 and September 21, during the year when the center of the earth lays in the plane of the sun and on these occasions the north-south axis of the earth is perpendicular to it equatorial plane and this aligning of the Sun's plane with Earth's plane and the positioning of the Earth's North-South axis and equatorial plane relative to the sun ensure that all areas of the earth receive equal hours of sunlight on these days,12 hours, FIG 7.3. These two times are referred to as the Spring (Vernal) Equinox, March 21 and the Fall (Autumnal) Equinox, September 21 and the relative positionings on these days also ensure that the hours of daylight are approximately also equal to the nocturnal hours, 12 hours.

Two important points to note about the relative positions of the Sun and Earth during these periods are as follows:

- The Earth is always above the plane of the Sun during its movement around its orbit from the time of the Autumnal Equinox to the Winter Solstice back to the Spring Equinox, September 21 through March 21(Fall through Winter to Spring), which would make the declination angle, D, in this period less than zero with a minimum of -23.45⁰.

- The Earth is always below the plane of the Sun during its movement around its orbit from the time of the Spring Equinox to the beginning of the Autumnal Equinox, Spring and Summer, and this make the declination angle, D, greater than zero with a maximum of + 23.45⁰, during this period.

Solar Angles

Angle of Declination

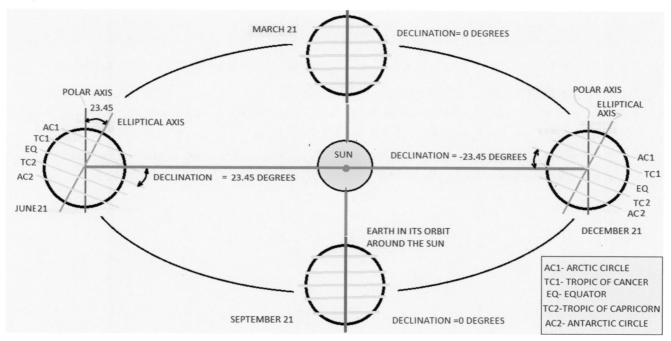

FIG. 7.4 Typical Diagram highlighting Maximum and Minimum Angles of Declination.

There are several angular relationships that significantly impact the amount of solar energy that may be collected at any particular point on the surface of the planet and these angles include:

- Solar angle of declination
- Hour Angle
- Zenith Angle
- Solar Angle of Altitude
- Solar Angle of Azimuth
- Latitude of the location
- Longitude of the location
- Solar angle of incidence

The declination angle is defined Kalogirou as follows:

- The solar angle of declination is the angular distance of the sun's rays north or south of the earth's equator.
- The angle formed between the sun's rays to the center of the earth and the earth's equatorial plane.
- The declination angle, D, varies throughout the year.
- A range of values for the declination angle are as follows, 0^0 at the Spring Equinox, $+ 23.45^0$ at Summer Solstice, 0^0 at Fall Equinox and $- 23.45^0$ at the Winter Solstice.
- The declination angle, for any day N of the year, may be calculated to give a value by utilizing an equation as follows, declination angle delta = 23.45 Sin{360/365(284+N)}

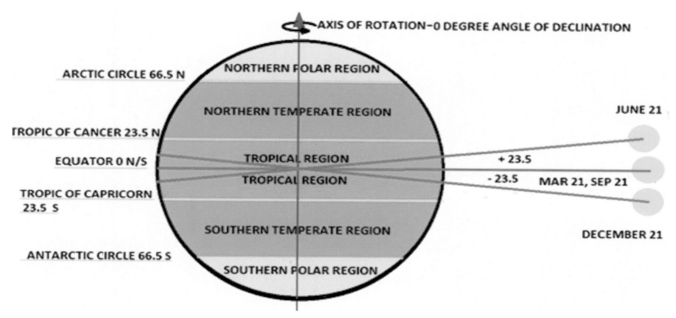

Fig 7.5 Typical Angles of Declination Relative to the Equator Highlighting the 3 Positions of the Sun

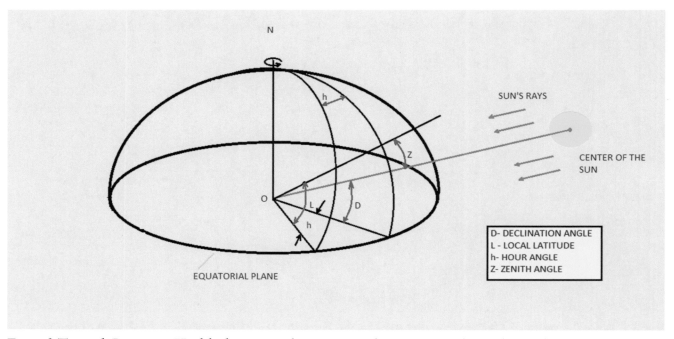

Fig 7.6 Typical Diagram Highlighting Declination Angle, Hour Angle And Local Latitude. Source: Kalogirou, Cyprus University Of Technology.

HOUR ANGLE

- The hour angle refers to a point on the earth's surface and is defined as the angle through which earth would turn to bring the meridian of the point directly under the sun.

- The hour angle at local solar noon is zero. Each degree of longitude is equal to 4 minutes based on the relationship of the earth making one complete revolution in 24 hours, 360/24 which gives 15 degrees per hour. This relationship may be expressed as h = +/- 0.25, where the plus sign is applied times afternoon and the negative sign to morning hours.

- The hour angle from a mathematical relationship with the Apparent Solar Time (AST), h = (AST-12)15.

- At local Solar Noon, AST = 12, h= 0, therefore the local standard time (LST), the time shown by clocks at local solar noon, may be obtained by the mathematical relationship LST =12 – ET+/- 4(SL-LL)

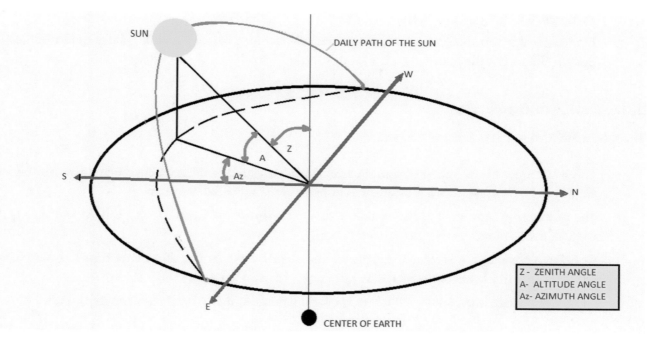

Fig. 7.7 Typical Diagram Highlighting The Position Of The Sun In The Sky.

ZENITH ANGLE, Z

The zenith angle is the angle Z shown on the diagram in Fig 7.7 and it is the angle between the rays of the sun and the vertical and this angle will change with the movement of the sun as it travels its daily path.

Altitude Angle, A

The altitude angle as follows:

- This is the angle between the Sun's rays and a horizontal plane as shown as angle A on the diagram above on Fig 7.7
- The altitude angle is related to the zenith angle Z as it the complimentary angle between the Sun's rays and the vertical.

- Z+A = pie/2 = 90^0
- The relationship can also be expressed mathematically as Sin(A) = Cos(Z) = Sin(L)Sin(D)+ Cos(L)Cos(D) Cos(h). Where h = the hour angle, L = local latitude

Solar Azimuth Angle, Az

The solar azimuth angle is defined as follows:

- This is the angle of the Sun's rays and the true south, measured in the horizontal plane towards the west and the movement westward is assigned as positive in the northern hemisphere.
- The relationship between the angles can be shown as Sin (Az) = Cos(D)Sin(h)/Cos(A) where D is the angle of declination.
- At Solar Noon the sun is located exactly on the meridian which includes the north-south line which would automatically make the solar azimuth angle, Az, equal to zero.
- The noon altitude is, A_n = 90-L+D where L = local latitude and D = declination angle

Latitude of Location

The latitude of a location is the angle between the lines joining that particular location to the center of the earth and the equatorial plane and the differences in the angles of latitude significantly impacts the amount of solar energy that may be collected at each location solar. The angle of latitude is measured from the equatorial plane with movements towards the north pole taken as positive and movement towards the South Pole taken as negative.

Longitude of Location

The lines of longitude on the globe are those great semicircles that join the North Pole to the South Pole and the measurement for longitude is usually take from a designated point, a Prime Meridian or the International Meridian, with movement to the east measured as positive and movements to the west as negative. The measure is usually in degrees with each degree representing 4 minutes of time or 15 degrees representing one hour.

Solar Angle of Incidence (I_a)

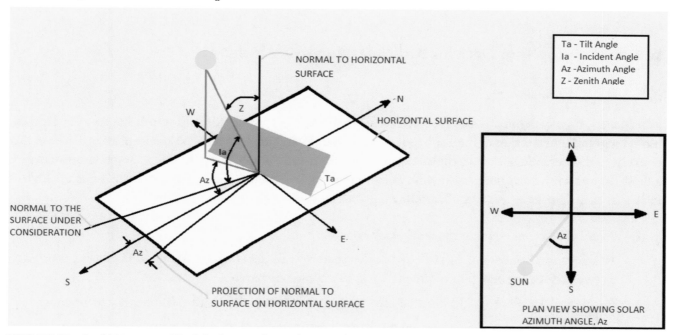

FIG 7.8 Typical Diagram Highlighting the Solar Angles on an Inclined Plane.

The angle of incidence, I_a, is the angle between the rays of the sun and the normal to a surface and for a horizontal plane the Angle of Incidence, I_a, and the Zenith Angle, Z, are the same,

I_a = Z.

The Angle of Incidence, I_a, changes at every instant of time and day and this angle must be maintained at a minimum to optimize the amount of solar energy to be collected. This angle is usually zero for a plane that is directly facing the Sun and at this angle the maximum amount of solar energy may be collected on a flat horizontal or inclined surface.

The relationship between the Angle of Incidence and the other solar angles may be represented as follows:

$$Cos(I_a) = Sin(L)Sin(D)Cos(T_a) - Cos(L)Sin(D)Sin(T_a)Cos(A_z) + Cos(L)Cos(D)Cos(h)Cos(T_a) + Sin(L)Cos(D)Cos(h)Sin(T_a)Cos(A_z) + Cos(D)Sin(h)Sin(T_a)Sin(A_z)$$

where T_a - Tilt Angle, A_z - Azimuth Angle, D – Declination Angle, L -Local Latitude, h – Hour Angle, I_a – Incidence Angle

Additional Important Data for Any Solar Energy Calculation

SOLAR NOON

Solar Noon is usually defined as that time of the day when the Sun's rays are directly perpendicular to a particular line of longitude and it occurs at the same time on each line of longitude and it will also take place one hour earlier for every 15 degrees of longitude to the east of any given line, such as a Primary Meridian, and one hour later for each 15 degrees west. Solar noon is the time when the sun crosses the meridian of the observer and is the highest point that the sun reaches each day. The following are also true about the Solar Noon:

- Solar Noon is not equal to the local clock noon
- To achieve desired correct Solar Noon or other solar time the local standard time (LST) must be corrected by two factors, the equation of time (ET) and longitude correction.
- Apparent Solar Time (AST) must be used to represent time of day in all solar energy calculations.
- Apparent Solar Time (AST) is based on the apparent angular motion of the sun across the sky.

EQUATION OF TIME

According to Kalogirou the earths orbital velocity varies throughout the year so the apparent solar time will vary along with the variation in velocity, however this variation is usually very slight in relation to the mean time kept by local timepieces which are running at uniform rates.

- This variation is usually called the equation of time (ET)
- ET = 9.87 Sin(2B) -7.53 Cos(B)- 1.5 Sin (B) {minutes}, where B = 360/364(N-81) and N= Day of the Year

Longitude Correction

Standard clock times are established relative to known meridians such as the Greenwich Standard Meridian, in England, which is at a longitude of zero degrees and this helps to establish the basis for longitude corrections.

- Longitude correction. The sun takes 4 minutes to traverse one degree longitude thus requiring a longitude correction factor = 4(Standard Longitude – Local Longitude) to be added to or subtracted from local clock times.

- If the location is east of the Standard Meridian the correction factor is added to the local clock time and if location is west of the standard meridian the factor is subtracted.

Apparent Solar Time

The Apparent Solar Time is based on the relative motion of the sun across the sky and when the sun passes through the meridian it is facing exactly south. The AST may be obtained from the following equation

AST = LST + ET +/_ 4(SL-LL), where LST = Local Standard Time, ET =Equation of Time. SL = Standard Longitude and LL = Local Longitude

East of the Greenwich Standard Meridian the sign is negative and west of it the sign is positive

Sun Rise and Set Times and Day Length

The sun rises and sets when the solar Altitude Angle is zero and from this the hour angle, h, at sunset may be found from by solving the equation for Altitude Angle when A= 0. The hour angle at local solar noon is zero.

The day length is twice the sunset hour since the solar noon is at the middle of the sunrise and sunset hours.

CHAPTER 8

Solar Radiation

Solar radiation is that great flow of electromagnetic waves of energy travelling at the speed of light, that emanates from that bright yellow star, called the Sun within our solar system which gave life to everything within the solar system and without which all life would cease. However, while very important to life on Earth, the Sun is only one of many such stars within the Milky Way Galaxy, the galaxy that Earth's solar system is part of and it is neither the largest nor the smallest of the stars in comparative sizes. This bright yellow star dwarfs all of the other planets that move around it paths called orbits, as it is much larger than all of the planets combined and it supplies all of the other planets within its reach with light, heat and gravitational push and pull that keeps them all within their designated orbits and at safe distances from each other. According to NASA, the Sun is approximately 864,000 miles (1,392,000 km) in diameter, has a surface temperature of $10,000^0$ F (5500^0 C), an inner core temperature of $27,000,000^0$ F ($15,000,000^0$ C) and it has a complex structure which consist of many layers and regions starting from a deep, dense, inner core and becoming less dense as the layers moving outward towards the outer layer of the Sun's atmospheric level, called the Corona which is many times hotter than the Sun's surface. The Sun is a giant, gaseous, spherical, self-gravitating body which consists mainly of hydrogen and it has very strong gravitational forces which create a very high pressure at its inner core and it is this very high pressure at the inner core that facilitates the process of nuclear fusion which turns hydrogen into helium. This nuclear fusion consumes a large portion of the Sun's mass and releases extremely large volumes of energy and electromagnetic radiation which flows outwards towards the planets in the solar system and probably further outward into the Milky Way Galaxy.

The energy created by nuclear fusion in the inner core of the Sun flows outwards to the surface of the sun, the photosphere, through a series of heat transfer processes inclusive of the convection, radiation, absorption, emissions and re-radiations. The photosphere, the starting point of all visible radiation that reaches the surface of the Earth, has a temperature of approximately $10,000^0$ F (5500^0 C/5777^0 K) and it absorbs and emits a wide continuous spectrum of radiation. The electromagnetic radiation emitted from the photosphere of the sun travels in highspeed waves moving outwards and these waves have different wavelengths, (the wave's crest to crest distances) with longwave and shortwave, with the crest-to-crest distances determining which waves are assigned as short wavelengths, 0.2- 3 micro-meters and crest to crest distances greater than this range are assigned as long wavelength. This short wavelength range, 0.2-3 micro-meters, covers parts of the UV range 0.01-0.4 micro-meters, the visible range 0.4-0.71 micro-meters, the solar range 0.3- 3 micro-meters and a small part of the Infrared range 0.8-1000 micro-meter. Generally, hot bodies with temperatures in the range of 5500^0 C emit radiation with short wavelengths as they are more intense and contain more intense radiation flows with greater energy, while cooler bodies will emit less intense, cooler radiation with longer wavelengths. Each flow of photons from the Sun normally consists of a mixture of short and long wavelength waves inclusive of UV, visible light and infrared radiation, each of which contains different levels of energy and will have different impacts upon the Earth and all Earth-bound systems, inclusive of solar energy collecting systems. In the field of solar energy, the short wavelength, 0.2-3 micro-meters, electromagnetic radiation is of the greatest importance as they contain the most energy and will have the greatest impact in both thermal and photovoltaic systems which includes hot water heaters, concentrated solar power systems, photovoltaic and concentrated photovoltaic systems.

Blackbody and Blackbody Radiation

The Sun is located in space at a distance of 93 million miles (150,000,000 km) from the Earth and the electromagnetic radiation, light and heat, from the Sun can only reach the surface of the Earth by the process of radiation, as only the process of radiation can transfer heat and light through the vacuum of space. Radiation, the process by which the Sun's energies reach the surface of the Earth, may be described as the process by which heat waves are given off by hot bodies and these heat waves may be absorbed, reflected or transmitted through a body at a lower temperature and through the boundaries of a vacuum as a body is not required for radiation heat transfer to take place, unlike the other two heat transfer methods, convection and conduction. Different bodies at different temperatures will give off different levels of heat waves, depending on the temperature of the bodies and the type of body, with the body that gives off the highest amount of heat waves at a particular temperature, being called a blackbody.

The definition of a blackbody states that "A blackbody is a body that can completely absorb radiation at any wavelength and completely emit radiation at any wavelength so that it emits more radiant power at each wavelength, than any other object at the same temperature" Zwinkels (2015). Blackbody radiation is also defined by Zwinkels (2015) as the radiant energy that is given off by a perfect blackbody whose range of heat wave distribution, spectral power distribution, is only determined by its own temperature. The concepts of a blackbody and blackbody radiation are critically important to all solar energy systems as all solar systems are dependent upon a perfect blackbody, the Sun, giving off radiant energy that can be absorbed by other blackbodies inclusive of enhanced absorber coils and plates in water heaters, absorbers and receivers in CSP systems and solar cells in PV and CPV systems. These two concepts, blackbody and blackbody radiation, are governed by several physical laws inclusive of the Planck Radiation Laws, Wiens Displacement Law and the Stefan-Boltzmann equation.

Planck Radiation Law

Electromagnetic radiation moves as a wave with wavelength of wl, that travels at the speed of light c =2.998x 10^8 meters per second in a vacuum or c/n in a material with a refractive index of n, and it has a frequency of f_{ew} that creates a relationship **c= wl. f_{ew}** where c = the speed of light, 2.998 x10^8 m/s, wl = wavelength of the electromagnetic wave and f_{ew} = the frequency of the occurrence of the electromagnetic wave.

According to Planck, electromagnetic radiation with a wavelength of wl is a flow of photons that has energy as provided in the following equation:

E = h. f_{ew} = h. c/wl, where E= photon energy, h = Planck's constant, 6.626 x10^{-34}

Planck also posited that thermodynamics and quantum mechanics dictates that a blackbody may be described by a law which states that "The wavelength distribution of radiation emitted from a blackbody with a temperature T is governed by the following equation, which is called Planck's Radiation Law:

G $_{bwl}$ = 2 pie hc²/wl⁵ (hc/ewlkT -1), where k = Boltzmann constant, 1.38x 10^{-23} J/K

The relationship between the wavelength and the temperature is governed by the equation

wl$_{max}$ = b/T where b is Wiens Displacement Constant = 2897.8 micro-meter K and T is the absolute temperature

The relationship above is known as Wien's Displacement Law, which states that the maximum wavelength is inversely proportional to the temperature, T, and this relationship is a derivative of Planck's law.

Another important relationship developed from Planck's Law, by Integration, is the Stefan-Boltzmann equation which states that the total blackbody radiation, G_b, is governed by the following equation:

G_b = S.T⁴ where S, sigma is the Stefan-Boltzmann constant, 5.670x10^{-8} W/m² K⁴
And T is the absolute temperature measured in Kelvins

Wien's Law and the Stefan Boltzmann equation highlights two significant facts that are fundamental and familiar about heated objects:

- Wien's displacement law highlights that the peak wavelength decreases with increasing temperature.
- The Stefan-Boltzmann equation highlights that the total radiated power increases with the temperature of the object.

Solar Constant

From the Stefan-Boltzmann equation the total radiative flux of the Sun can be calculated by the following equation:

G_{bt} = S.T⁴. 4.pie. r², where r is the radius of the Sun and 4.pie.r² is the surface area of the Sun.

The radiative flux from the Sun to distant planets will also be given by a similar equation:

G_{btdp} = S.T⁴. 4. pie. r² / 4. pie. l² = S.T⁴(r²/l²), where l = to the distance between the sun and the planet.

This equation when applied to the relationship between the Sun and the Earth will produce what is known as the Earth's solar constant which is the radiative flux just outside of the Earth's atmosphere and this is G_{sc} = **1367 W/m²** which varies relative to the position of the Earth in its orbit around the Sun. This value varies from 1412 W/m² at the beginning of July to 1322 W/m² at the end of the year due to the Earth's elliptical orbit around the Sun and the solar constant for each planet may also be derived in the same manner.

Categorization of Radiation

The Sun emits solar electromagnetic radiation with wavelengths in a range of 0.1- 4 micro-meters and this range consists of several bands of different wavelengths, these bands have been classified into several categories inclusive of the different wavelength bands, their properties and the different areas of application, with categories are as follows:

- Ultraviolet 0.1 – 0.4 micro-meter (Short-wave)
- Visible Light 0.40 - 0.71 micro-meter (Short-wave)
- Infrared 0.71 – 4 micro-meters (Short-wave to long-wave)

- Near Infrared (NIR) 0.71-1.5 micro-meter (Short to long-wave)
- Far Infrared (FIR) 1.5 – 4 micro-meters (Short to long-wave)

The wavelengths ranging from 0.1 – 0.71 micro-meters, which covers the Ultraviolet to the Visible range, are categorized as short-wave and the range above this as long-wave and this categorization into short-wave and long-wave is very important in the design of most solar systems as both sets of wavelength, short-wave and long-wave, have different thermal and optical properties and will impact the collection of incident solar radiation, as short-wave radiation have far more energy than long-wave radiation and the material required to collect short-wave and long-wave solar radiation may be quite different.

Atmospheric Attenuation

The flow of solar radiation from the Sun's surface, photosphere, to the Earth's surface can be hampered and impeded by many elements inclusive of space dust, particles, gases and bodies that have the capacity to partially or wholly block the flow of the Sun's radiation, as happens when an eclipse occurs and the flow of electromagnetic radiation from the Sun is blocked by the movement of celestial bodies between the Sun and the Earth. This type of interference also occurs within the atmosphere of the Earth which contains extremely large volumes of dust particles, water vapor and gases, all of which are a permanent part of the atmosphere with varying density dependent upon weather conditions, natural events and anthropogenic activities. Weather conditions will see an increase or decrease in the amount of water vapor in the atmosphere, while volcanic activities will create a significant increase in the particles and gas concentration in the atmosphere and events such as sand storms, the movement of Sahara dust, will likewise greatly increase the concentration of dust particles in the atmosphere. Natural events alone, however, is not responsible for the excessive amount of particulate matter and gases that currently resides within the atmosphere, as transportation systems, manufacturing and other industrial activities emits millions of metric tonnes of particulate matter and gases into the atmosphere each day. This large volume of natural and anthropogenic dust particles, water vapor, particulate matter and gases all flow across the path of incident solar radiation on its way to the surface of the Earth and this flow significantly impacts the natural direction of flow and the energy contained within the solar radiation that eventually reaches the surface of the Earth. These impacts could cause significant changes to the direction of flow, the level of energy contained in the photons that reach the Earth and the final destination of the photons that come in contact with the molecules and particles, thereby causing a reduction in the capacity of the solar radiation to perform all of it natural function as it gives up some of its energy to these billions of particles. The ability of particles in the atmosphere to negatively impact the natural energies of incident solar radiation is referred to as atmospheric attenuation, which mainly occurs in two forms, scattering and absorption.

Scattering

Scattering happens when the solar radiation interacts with air molecules, water vapor and particulate matter in the atmosphere and the extent of the scattering achieved is a function of several factors inclusive of the wavelength of the solar radiation relative to the size of the particle, the concentration of the particles in the atmosphere and the total mass of the air through which the solar radiation must pass. After impact with molecules or particles, light may be scattered in several patterns inclusive of some of the more prominent patterns such as the Rayleigh Scattering, the Mie Scattering and the Raman Scattering and of these three different patterns, the Rayleigh Scattering is considered to be the most significant. In the Rayleigh Scattering process short-wave wavelengths of light at the blue end of the light spectrum, with wavelengths smaller than 0.6 micro-meter, are scattered off air molecules that are about a tenth of the size of the wavelength of the light and the scattered light is sent up into space to produce the blue colour of the sky. This Rayleigh Scattering process provides an understanding of why the sky is blue during daytime hours, the yellow colour of the sun and the reddening of the sky as night approaches.

In the Rayleigh Scattering pattern, the photons of light retain their energy at the same levels and thus is usually called "elastic scattering", however, this is not the case in all of the scattering processes, as in the Raman Scattering the photons will lose or gain energy based on the excitation of some vibrational mode of the molecule.

Absorption

The absorption of solar radiation occurs across a wide range of the electromagnetic spectrum inclusive of the UV and IR wavelength ranges and the electromagnetic radiation at these wavelengths tend to impact some of the gases with significant outcomes that affect the climatic conditions on the Earth. The UV radiation tend to impact the Ozone in the stratosphere and the absorption of UV by Ozone is of great importance to life on Earth as this process prevents excessive amounts of UV radiation from reaching the Earth's surface, as excess UV radiation at ground level can be quite harmful to all forms of animal, plant and human life. The IR wavelength radiation interacts with water vapor and carbon dioxide in the atmosphere and in this interaction the IR radiation, in cases of water vapor and carbon dioxide, is absorbed, is converted to heat and this heat is later emitted by the carbon and water vapor molecules as long-wave radiation which travels back to the earth, impacts clouds and some of it travels back out into space where all three components can both positively and negatively impact life on Earth.

The absorption of IR by water vapor and carbon dioxide and its reradiation as longwave heat has long been one of the most important processes that has helped to foster life on Earth, as the reradiated heat has worked together with the other elements of the atmosphere to warm the globe to an average liveable temperature of 59^0 F, well above the mathematically projected temperature of 0^0 F, exemplifying the global warming effects of these gases. On the negative side the presence in the atmosphere of excess amounts of these global warming gases such as carbon dioxide and water vapor, which retain their natural ability to absorb IR radiation and reradiate it as heat, will cause excess heating of the Earth, inclusive of land, surface water and the atmosphere and this excess heating can lead to great changes in climatic conditions all around the globe. The envisioned possible changes that could result from the excessive heating includes rise in sea levels due to expansion and increased inflow from the melting of permanent ice, ice caps and glaciers, excessive swings in temperature and precipitation levels and extreme levels of wind activities inclusive of more frequent massive hurricanes and tornadoes. The realization of these changes could cause great loss of lives and would consequently force major changes in traditional and existing human living patterns which were developed over millennia, if the human species and many species of plants and animals are to survive. This necessary readjustments in living patterns to ensure the survival of the human race and many species of flora and fauna under the conditions created by climate change could take millennia to realize and, in the interim, there could be a great dying off of humans and many other species, especially those living in coastal regions of the globe, before the required long-term solutions are formulated and implemented.

Components of Radiation

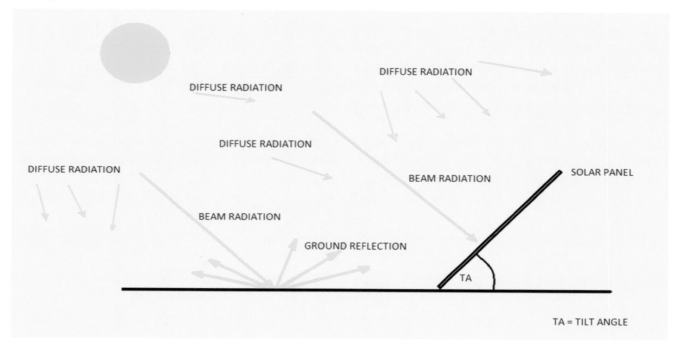

Fig. 8.1 Typical Diagram of Solar Radiation Components. Source: Solar Radiation Theory

The attenuation of the incoming electromagnetic radiation produced by scattering and absorption has a significant impact on the solar radiation that reaches the surface of the earth as the attenuation processes have caused the incoming electromagnetic radiation to be broken up into two main components, direct beam radiation and diffused radiation, both of which contain significant amounts of energy that can be utilized to produce electrical energy and heat in solar energy systems.

Diffused Radiation

Diffused radiation is produced from the portion of the incoming radiation that was scattered by air molecules, dust particles, water vapor, pollutants, and clouds, as described in any of the three main scattering theories and patterns, Rayleigh, Mie, and Raman. This scattered radiation tends to move in two major directions, up into the atmosphere towards space where some is reflected back towards the surface of the Earth and down towards the surface of the earth, as diffused radiation. This diffused radiation is not well defined and is usually said to originate from the whole dome of the sky but it is uniformly distributed in all directions and usually only forms a portion of the incoming electromagnetic radiation as on a bright sunny day it accounts for between 10-20 percent of the incoming solar radiation. However, on a cloudy day when there is no bright sunshine nearly all of the incoming radiation is diffused and similarly on days when there are high levels of dust and other particles in the atmosphere diffused radiation also dominates. The highest levels of diffuse radiation usually occur during the times when there are thin broken clouds overhead and the amount of diffused radiation produced on a partially clouded day is substantially higher than that produced in a clear sky and from a typically overcast sky.

Diffused radiation also has a distinct spectrum from that of direct beam radiation, as diffused radiation is created by the impact of short-wave radiation only, 0.6 micro-meter in the blue range, with particles, air molecules and gases, while direct beam contains the whole light spectrum.

Direct Beam Radiation

Direct beam radiation is the electromagnetic radiation which travels directly from the Sun to the Earth unimpeded as it passed through the atmosphere without being scattered or absorbed and this direct beam radiation is a vector quantity having both a magnitude and a direction and it travels in a straight line from the Sun and contains far more energy than diffuse radiation. A comparison of Direct Beam Radiation and Diffused Radiation is shown in the table below.

ITEM	DIRECT BEAM RADIATION	DIFFUSE RADIATION
1	Straight lines from the sun to the Earth	Scattered lighting reach the Earth
2	Vector quantity -magnitude and direction	Non-directional
3	High overhead sun produces 85 %	High overhead produces 15%
4	Very intense	Less intense
5	Intensity declines down to 60% with declining sun each day	Intensity increases up to 40% with declining sun at 10^0 above horizon each day.
6	Solar panel must be inclined to ensure maximum collection.	Solar panel must be horizontal to ensure maximum collection.
7	Intensity varies with latitude and climate, intensity declines at high latitudes	Intensity varies with latitude and climate, intensity increases at high latitudes
8	Intensity decreases down to zero with cloud cover.	Intensity increases up to 100% with cloud cover
9	Intensity decreases in winter	Intensity increases in winter.
10	Produces more heat and electricity per panel	Produces less heat and electricity per panel

Table 8.1 Direct Beam Radiation V Diffused Radiation.

Direct beam radiation is currently the main factor and driving force in the design of all solar energy systems, PV and thermal, however, this status will be changing in the near future as the new solar energy technologies are now being designed to utilize diffused radiation in PV energy systems as well.

Reflected Radiation

Consideration must also be given to the impact of electromagnetic radiation that has reached the surface of the Earth and the portion of it that is reflected back into the atmosphere, onto buildings, onto solar panels, similar man-made systems, natural systems and then into space. This capacity of the ground to reflect solar radiation is a function of the albedo of the particular soil or ground cover inclusive of grass, asphalt, concrete and even snow, which reflects more radiation than most other ground covers. This ground reflected radiation also has significant energy value that can be absorbed by ground mounted solar energy collectors and must therefore be taken into consideration when designing these systems.

Air Mass

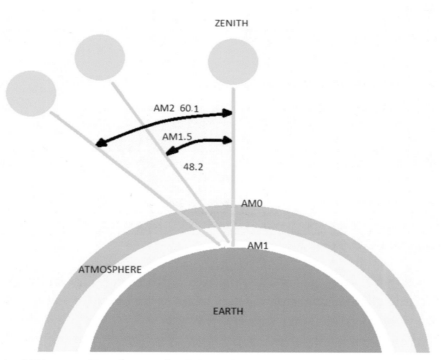

Fig 8.2 Typical Diagram Displaying Air Mass, Am0, Am1, Am1.5 & Am2. Source: Laser Focus World.

The attenuation of Solar radiation is also a function of the distance that the radiation has to travel in space and in the atmosphere before it reaches the surface of the earth, with longer distances ensuring more interaction with air molecules and particles. The distance to be travelled and the amount of air molecules and particles in the atmosphere varies with several factors inclusive of ozone layer condition, earth-sun distance, solar rotation, the time of the year and even the time of the day, as the longest distance travelled each day is when the sun is over the horizon in the evening hours. The comparison of the long and short distances that the solar radiation must travel in a day has led to the formulation of a quantity called the Air Mass which is used in the solar energy calculations and one definition of the air mass is as followed "Air mass is the ratio of the atmospheric mass through which the radiation passes from the Sun's current position in the sky to the mass that it would pass through if the Sun were at the zenith (directly overhead)" Widen and Munkhammer (2019).

The Air Mass changes with the daily movement of the Sun across the sky, the seasonal positions of the Sun as shown in Fig 7.2, which highlights the highest and lowest points of the sun in the sky, the movement of the Earth around the sun and the rotation of the Sun on its axis.

Solar Energy Availability

The sun's energy reaches every place on the surface of planet Earth but not every location on the surface of the planet receives the same amount of direct sunlight for 12 hours, every day, every month of the year, however, there are locations that receive the sun's full energy every day for 12 hours, every day of the year except during periods of extended cloud cover and precipitation, such as in periods of hurricanes, dust storms and similar weather events. There are many locations in both the Northern and Southern Hemispheres that receive more than 12 hours and up to 24 hours of sunlight each day during certain times of the year. There are several locations where the sun's full energy would only be available for several months of the year based on the movement of the sun, the tilted axis of the Earth, the rotation of the Earth around the Sun and the latitude of the location and this can be seen very clearly in both the Arctic and Antarctic regions of the planet, above the 66.5 N and below the 66.5 S positions, that receive full solar radiation for only six months of the year and partial radiation for the other six months of the year, especially at the two extremes of the poles. The diagrams below highlight the sunlight available at both the North and South Poles during the year.

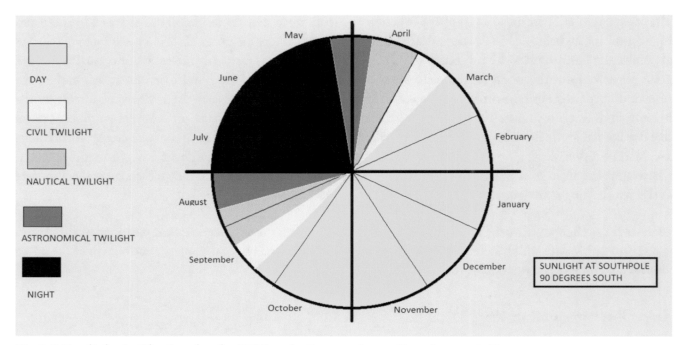

Fig 8.3 Sunlight At The Southpole, 90° South. Source: Australian Antarctic Program

At the South Pole the conditions as demonstrated in Fig 8.3 apply 24 hours per day, however, for locations further north, such as 80° South, the hours of daylight, civil twilight, nautical twilight, astronomical twilight and darkness may vary hourly, daily and monthly as you move further north up to the Antarctic Circle at latitude 66.5° S, after which conditions again change significantly. The basic definition of the various partial lighting conditions are as follows:

- Civil Twilight. This the condition after sunset and before sunrise when the centre of the Sun is approximately 6° below the horizon and the brightest stars and terrestrial objects are still visible.

- Nautical Twilight. This is the condition that occurs when the Sun is between 6° – 12° below the horizon and in this condition only the general outline of ground objects and the horizon may be visible.

- Astrological Twilight. This is the condition that occurs when the Sun is between 12° -18° below the horizon and in this condition, there is no colour in the sky and the horizon is not visible.

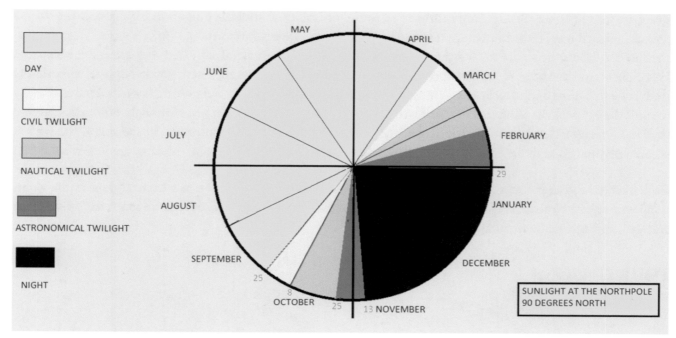

Fig. 8.4 Sunlight At The Northpole 90° North.

At the North Pole the conditions as demonstrated in Fig 8.4 also apply 24 hours per day, however, for locations further south, such as 70° North, the hours of daylight, civil twilight, nautical twilight, astronomical twilight and darkness may vary, hourly, daily and month by month as you move further south until the Arctic Circle, 66.5° N, after which conditions again vary greatly. The basic definitions of the various partial lighting conditions are as given above.

A comparison of the two diagrams, FIG 8.3 and 8.4., will clearly highlight that the periods of total darkness occur at two different times, May to July in the South Pole and November to January in the North Pole, the corresponding periods of winter at both poles. These diagrams along with the diagrams on Fig 7.3 also highlights the great differences in the variations in the availability of sunlight at different locations around the globe due to the direction of the axis of the earth relative to the sun, the orbital rotation around the Sun and the latitude of the location. The variations at the two poles are the extremes of the variability of available sunlight, however, these variations clearly point to variability of available sunlight at other locations going from the extremes position, North and South Poles, with each change in latitude, time of day, time of month and time of year.

There are four different times during each year when the variability of available sunlight is highlighted, the Winter Solstice, the Summer Solstice and the Spring and Fall Equinoxes. During the Northern Winter Solstice, December 21, the Sun is at its lowest point in the sky, the axis of the Earth is at its farthest point away from the Sun and the Southern Hemisphere receives a larger volume of direct sunlight during the daytime than the Northern Hemisphere, while during the Northern Summer Solstice, June 21, The Sun is at its highest point in the sky and the earth's axis is tilted maximally towards the sun with the result that the Northern Hemisphere receives more direct rays of sunlight during the daytime than the Southern Hemisphere. At the time of the equinoxes, March 21 and September 21, the axis of the earth is vertical, perpendicular to the Sun, the northern and southern hemispheres will receive the equal amount of sunlight.

Each of these dates represent a major turning point in weather conditions in the temperate regions of the globe, temperate region North 23.5N - 66.5 N and temperate region South 23.5 -66.5 S, that are impacted by the "four seasons", spring, summer, fall and winter.

FOUR SEASONS

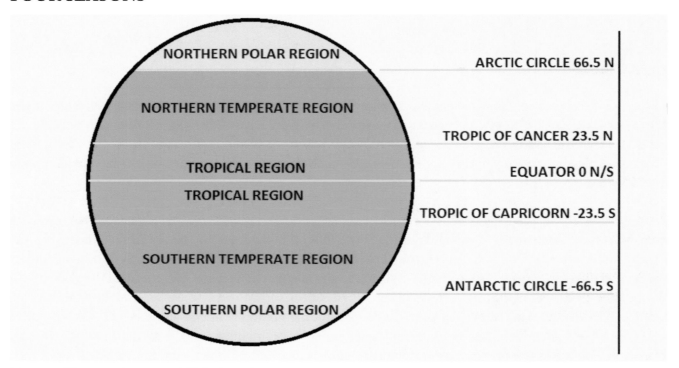

Fig. 8.5 Climatic Regions Of The Earth.

The conditions referred to as the four seasons are cyclical weather conditions in the temperate regions of the Earth and these regions lay between the very cold polar regions and the hot tropical regions north and south of the Equator as demonstrated in FIG 8.5 above. These weather conditions typically correspond to the relative movement of the Earth, the Sun, the relationship of the earth axis relative to the Sun and the continuous movement of both bodies relative to each other. These four seasons are called Spring, Summer, Autumn and Winter and each season occurs at one particular time of the year, when the Earth is in a particular section of the Earth's orbit around the Sun as highlighted on Fig 8.6 below.

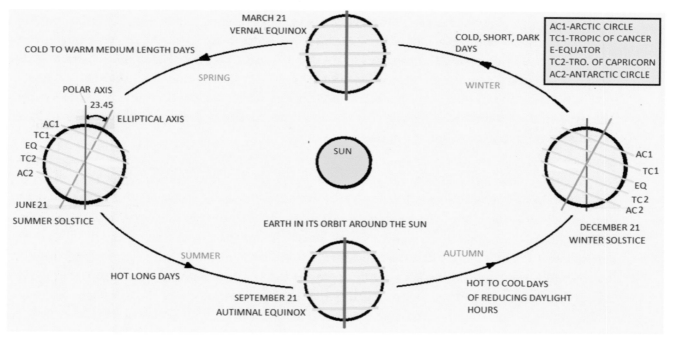

Fig 8.6 Typical Diagram Highlighting The Four Sesaons Relative To The Earth's Movement Around The Sun, As Seen In The Northern Hemisphere.

- **Spring**. March 21, Vernal or Spring Equinox, represents the beginning of the growing and planting season, Spring, and at this point the axis of the Earth is vertical and is neither pointed towards or away from the Sun but it is halfway, 90^0 away from the Sun, along the track to reaching the position when it is directly pointed at the sun, at 23.5^0. On this day, the Spring Equinox, there are an equal number of daylight and night time hours following which the length of the days, number of hours of daylight, will increase incrementally and this will allow more time for the collection and utilization of solar energy by plants, animals and solar energy collecting systems, many plants and animals take a very long sleep (hibernate) during the winter months and only come awake at the beginning of Spring. The season of Spring runs from March 21 to June 20 each year and the weather in Spring goes from cold, snow and ice, to warm with the melting of snow and ice and finally hot with the approach of the summer season. The melting of ice and snow and the liquid precipitation which comes with the changing seasons, when combined could cause significant flooding in the flatland areas due to the flow of melted snow and ice in the hills and mountains. The timely melting of the snow and ice has for many millennia been very critical for agriculture, the planting and growing of crops, however, if the atmospheric and surface temperatures get higher than normal too quickly, due to climate change or any other climatic conditions, the lowland regions will become flooded, with the excess water doing damage to crops, animals, roads and infrastructure, after which it flows away to the rivers, streams, ponds and lakes where it cause further damage to aquatic plants and animals which also form a part of the human food chain before it flows away to the sea where it's direct benefits are lost to agriculture and anthropogenic water collection systems. The speed of the flood waters over land may also limit the amount of these water that may be collected in natural cisterns and underground aquifers.

- **Summer**. June 21, Summer Solstice is the longest day of the year and represents the beginning of the hottest and brightest period of the year, Summer, a period of time that facilitates the growth and maturation of all plants inclusive of long-term crops and provides an opportunity for the growth and development of animals also. The Earth at this time is at the point at which its axis, at 23.5⁰, is directly pointed towards the Sun and the Earth receives the maximum amount of solar energy in the northern hemisphere. After this longest day, the number of hours of daylight, while still relatively long, begins to incrementally reduce and this automatically begins to reduce the amount of solar energy that may be collected on the days thereafter. Summer usually starts hot and become hotter and drier right through to the beginning of Autumn when the temperatures begin to cool down with the aid of thunder storms and tropical hurricanes, some of which usually traverse the Atlantic Ocean starting off the coast of Africa, moving through the Caribbean, the Americas and sometimes eventually die out in the upper Atlantic Ocean or moving into some European land masses inclusive of Ireland and the UK. Some hurricanes also originate in the Pacific Ocean and move on the western shores of the Americas and other land masses in the Pacific Ocean. Both sets of hurricanes and thunder storms bring with them significant amount of precipitations which cools the Earth and surrounding systems but they usually also cause significant damage to all facets of life, human and the environment, as hurricanes and thunderstorms tend to be a combination of heavy precipitation that can cause flooding and high winds that can destroy trees and plants, electrical infrastructure and building of every type.

- **Autumn.** September 21, Autumnal or Fall Equinox, represents the beginning of the repeating season when the leaves of deciduous trees start ripening before falling off the trees, Fall, and at this point in time the axis of the Earth is halfway,90^0 away from the Sun, along the track of reaching the position when it is furthest away from the Sun. During this period the length of the days will be incrementally reducing leading to the shortest day of the year December 21 and the beginning of the very cold season when the days are cold, dark and short. During this season, Fall, the weather transitions from the heat of summer to the cold of Winter, when most natural life slows down.

- **Winter.** December 21, Winter Solstice is the shortest day of the year, represents the end of the Autumn or Fall period and the beginning of the period when some of the natural systems of the earth, inclusive of trees and some animals, slows down and go into a long sleep. At this point in its movement around the Sun, the axis of the earth is at its furthest point away from the sun creating the coldest weather in the Northern Hemisphere. On this day the number of daylight hours are now reduced to its lowest level in the year and after the 21st, daylight hours will begin to incrementally increase giving more hours of daylight and more opportunity to collect solar energy, where daylight hour are still available in the Northern Hemisphere. Winters in both the Southern and Northern Hemispheres are usually very cold, falling to temperature in the negative range of thermometers, with shorter daylight hour, with precipitation in the form of snow, sleet and ice, which remain in the mountains, on the ground or on trees and buildings until the weather gets warmer and this pattern of short daylight hours, low temperatures and solid precipitations usually continues until the Spring season of each year.

The conditions of the four seasons Spring, Summer, Autumn and Winter, remain the same but they do not occur at the same time in the Northern and Southern Hemispheres as outlined above, with the information above relating only to the Northern Hemisphere. In the Southern Hemisphere the pattern is as follows:

- **Spring.** Spring occurs between September 21- December 20, autumn in the north, with September 21 as the Spring Equinox, and has the same characteristics as Spring in the north, thawing and melting of snow and ice, high flows of melted water, heavy precipitations, the awakening of plant and animals and the beginning of new life.

- **Summer.** Summer occurs between December 21- March 20, with December 21 as the Summer Solstice the longest day of the year, the Winter Solstice in the North, and has the same characteristics as summer in the north, long, hot, bright days with the highest levels of insolation, higher surface temperatures and greater solar energy potential, both photovoltaic and thermal energy.

- **Autumn**. Autumn occurs between March 21- June 20, with March 21 as the Fall Equinox, Spring Equinox in the north, and has the same characteristics as Fall in the north, long bright days transitioning into shorter colder day, the maturation period for many crops, the leaves of trees ripening before falling off and some reduction in the amount of solar energy available to create PV and thermal energy.

- **Winter**. Winter occurs between, June 21- September 20, with June 21 as the Winter Solstice the shortest day of the year, has the same characteristics as winter in the north, short, cold days where natural life slows down and there is a small solar energy potential.

The amount of sunlight received in the Northern and Southern Hemispheres varies going from a high in the summer time, Summer Solstice, to a low in the winter time, Winter Solstice, passing through the equilibrium points during the two Equinoxes, with each hemisphere having their own Summer and Winter Solstices. Generally, the Sun's movements across the sky along with the local climatic and weather conditions, time of the year, altitude and landforms affect the solar resources in different geographical locations, with the geographical locations including the general latitude of the location. The amount of solar energy available is a function of available solar insolation which varies significantly with changes in latitudes and coastal plains regions having much greater levels of insolation than in mountainous regions.

The variability in the availability of solar energy at different locations around the planet determines the amount of solar energy that may be collected and utilized for photovoltaics and solar thermal energy systems everywhere on the surface of the earth. Table 8.2 below highlights the variability in the number of daylight hours that is available in 10 regions of the planet, however, daylight hours alone does not determine the amount of solar energy that may collected as it has been established that solar system are less effective in the Northern Hemisphere, as the insolation, solar radiation Kwh/m^2/day or Kwh/m^2/year is reduced with any increase in latitude, north and south.

LOCATION	LATITUDE	DAYLIGHT HOURS (LOW)	DAYLIGHT HOURS (HIGH)	DAYLIGHT HOURS AVERAGE
ANCHORAGE	61.20 N	6 (JAN & DEC)	18 (JUNE)	12
LONDON	51.50 N	8 (JAN & DEC)	16 (JUNE)	12
TOKYO	35.67 N	10 (JAN, NOV &DEC)	14 (JUNE)	12
HOUSTON	29.76 N	10 (JAN&DEC) 11 (NOV)	14(MAY-JULY)	12
CARACAS	10.48 N	11 (JAN -APRIL) & (AUG -DEC)	13 (MAY-JULY)	12
NAIROBI	1.290 N	12	12	12
RIO de JANIERO	22.90 S	11 (MAY-AUG)	13 (DEC)	12
SIDNEY	33.86 S	10 (MAY-JULY)	14(JAN, NOV&DEC)	12
CAPE TOWN	33.92 S	10 (MAY-JULY)	14(JAN, NOV&DEC)	12
BUENOS AIRES	34.60 S	1O (MAY-JULY)	14(JAN, NOV&DEC)	12

Table 8.2. Daylight Hours Across Latitudes Around The Globe. Source: Usra(Nasa)

Solar Insolation

Solar insolation refers to the quantity of solar radiation that reaches the surface of the Earth and this quantity is only a portion of the solar radiation that reaches the top of the Earth's atmosphere, with the solar insolation is typically expressed as units of energy usage per meter squared per day or per year as follows, Kilowatt-hour per meter squared per day and Kilowatt-hour per meter squared per year. The amount of solar insolation that reaches the surface of the Earth is a function of several factors inclusive of the Sun's movement across the sky, landforms inclusive of mountains and plains, atmospheric conditions inclusive of clouds and particulate matter, Sun Angle, Air Mass, the Angle of Insolation, the duration of insolation, the latitude of the location and the season of the year.

The Sun's position in the sky is a function of its movement from east to west each day and its movement from south to north over the period of the year with the Sun reaching its lowest position in the sky during the Winter Solstice, 24.5^0, its highest position during the Summer Solstice, 71.5^0 and its middle position during the Equinoxes, 48^0, positions which are all a function of the movements of the Earth relative to the Sun.

Landforms, mountains, plains, rivers and lakes also affect the insolation, with areas in the coastal plains tending to have much higher levels of insolation than mountainous regions of the same latitude as the temperatures in the mountainous regions tend to be much lower, due to altitude, than on the plains. Atmospheric conditions inclusive of cloud cover, gas molecules and particulate matter will impede the path of the direct solar radiation from the Sun and thereby ensure that less direct solar energy reaches the surface of the Earth and these atmospheric conditions, as earlier mentioned will scatter and absorb incoming solar radiation and thereby reducing the intensity of the incoming solar radiation.

The Sun's angle or angle of insolation is of great importance as this angle determines the distance that the radiation must ravel before it reaches the surface of the Earth, with the highest point it reaches in the sky, its Zenith, creating the shortest direct path where it loses the lowest amount of energy due to scattering and absorption, relative to the other angular positions that it travels through when it comes over the horizons early in the morning moving up to it zenith and when it declines and goes over the horizon in the evenings. The ratio of the different distances travelled by the sunlight, from its rising position through to zenith and decline is called the Air Mass which is described as the ratio of the Zenith height and the length of travel of other positions during the process of the daily crossing of the sky and usually has values such as AM 0, AM 1, AM 1.5, AM 2 and AM 3. The intensity of the insolation will be greatest at AM1when the Sun is at its Zenith and where the angle of insolation is closest to 90^0 and this angle of insolation varies through the day increasing in the morning up to Solar Noon and decreasing in the afternoon into evening when the Sun goes below the horizon.

The duration of the insolation will be function of the Sun's position in the sky, 24.5^0 – December 21, 48^0 March 21 and September 21 or 71.5^0 -June 21, each of which determines the Sun's travel path across the sky and the number of daylight hours, with the shortest daylight times occurring during the fall and winter periods and the longest daylight times occurring during the Spring and Summer periods. Insolation levels will therefore be greater from Spring to the end of Summer than from Autumn to the end of Winter in the temperate regions of the Earth and remain approximately the same in the tropical regions. Solar insolation also varies throughout each day with the lowest levels occurring at sunrise and sunset and the highest levels occurring at solar noon each day as demonstrated in the graph on Fig 8.7 below.

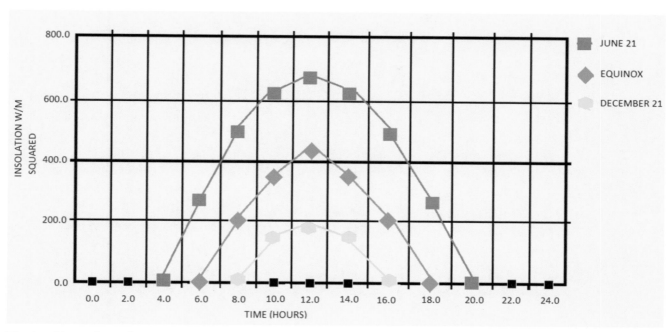

Fig 8.7 Typical Daily Variation In Insolation Levels.

The daily variation in insolation is location specific and the graph in FIG 8.7 is intended for general information only as things like the amount of daylight hours available for the collection of solar insolation could be different in each location and may vary based on the latitude, season and time of the year and climatic conditions. The constants for each location are that the highest levels of insolation will occur during the summer when the days are longer and the sun is highest in the sky and the lowest levels will occur during winter when the days are shortest and the sun is lowest in the sky. As previously stated, summer and winter occur at different times in the Northern and Southern Hemispheres with summer in the northern hemisphere occurring at the same time as winter in the southern hemisphere, June 21- September 20 and winter in the northern hemisphere occurs as the same time as summer in the southern hemisphere, December 21 -March 20.

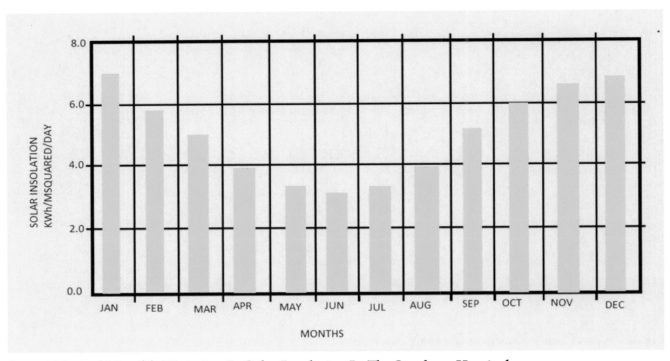

Fig 8.8 Typical Monthly Variation In Solar Insolation In The Southern Hemisphere.

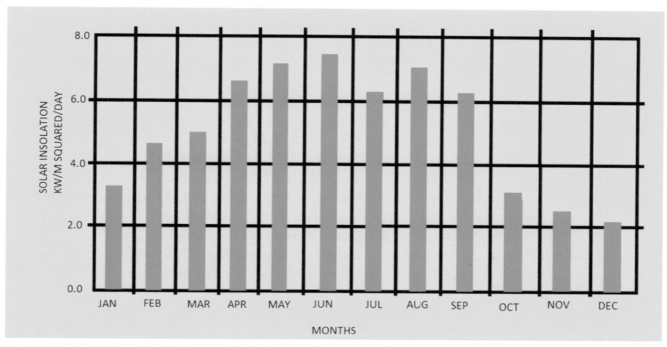

Fig 8.9 Typical Monthly Variations In Solar Insolation For The Northern Hemisphere.

Figs 8.8 and 8.9 are only the general shape of the graph that will be formed based on the monthly variation in solar insolation in the Northern and Southern Hemispheres as the insolation in each hemisphere will also be impacted by latitude, weather conditions and landforms. Each of these conditions will determine the true insolation for each month and impact the exact shape of the graph in each location, except for two things, both hemispheres will have high and low points for insolation, the high point in the northern hemisphere occurs in summer which runs from June 21- September 21 and the low point occurs in winter which runs from December 21 – March 21, While in the Southern Hemisphere the times of the high and low points are reversed with summer occurring from December 21- March 21 and winter occurring from June 21-September 21. Based on available information, a monthly insolation graph can be produced for each location based on their latitude and other location specific conditions and these graphs and the information contained can facilitate the work of solar energy system planners and designers.

Latitude of Location

The latitude of location is of the greatest importance to the amount of insolation a location receives as any increase in latitude away from the position of the Equator will ensure a reduction in the amount of insolation that the location will receive. The highest level of insolation is usually received when the Sun is directly overhead and the Sun is only directly overhead between latitudes 23.5^0 South and 23.5^0 North which occurs at different times during the year. At the time of Northern Hemisphere's Winter Solstice, the Sun is directly overhead over the Tropic of Capricorn and at time of the Northern Hemisphere Summer Solstice the Sun is directly overhead at the Tropic of Cancer and these two positions cover the tropical region of the globe and this physical reality clearly indicates that regions south of latitude 23.5^0 South and north of latitude 23.5^0 North will only receive sunlight at angle less than 90^0. This sunlight at less than 90^0 typically produces less insolation and this reduction in insolation is a function of the increasing distance that the solar radiation must travel to reach the surface of the Earth at the higher latitudes. In Northern Hemisphere the Sun is never directly overhead above the Tropic of Cancer, 23.5^0 N, and the solar insolation tends to reduce as the latitude increases moving towards the North Pole. At these higher latitudes the solar insolation reaches the surface of the Earth at less than 90^0, the optimum angle, which only occurs in and around the tropical region, and the further north that the sunlight has to travel the insolation angle will be further reduced and the amount of insolation received becomes commensurately smaller. At latitude higher than 66.5^0 N in the Arctic Polar Region, there is usually a significant amount of sunlight however, the solar insolation is almost parallel to the surface of the Earth and has relatively low insolation impact. Similarly, in the Southern Hemisphere the furthest south that the Sun will be directly overhead is the Tropic of Capricorn and solar insolation will also decrease as latitudes increase going towards the South Pole and the angle at which the solar insolation reaches the surface of the Earth will continue to decrease until it also becomes almost parallel to the surface of the Earth and has relatively low insolation impact, which indicated that these regions receive a lot of diffused sunlight.

The impact of latitude on insolation in locations in both the Southern and Northern Hemispheres are as demonstrated in the table below and the graph that follows.

LOCATION	LATITUDE	AVERAGE IRRADIANCE Kwh/m²/day		
		DNI	GHI	PVP
ANCHORAGE	61.20 N	no data	no data	No data
LONDON	51.50 N	2.4	2.85	2.8
TOKYO	35.67 N	3.4	3.9	3.8
HOUSTON	29.76 N	4.5	4.7	4.2
CARACASS	10.48 N	4.0	5.0	4.2
NAIROBI	1.290 N	4.0	5.3	4.2
RIO de JANIERO	22.90 S	4.0	4.8	4.0
SIDNEY	33.86 S	5.0	4.6	4.2
CAPE TOWN	33.92 S	6.0	5.2	4.8
BUENOS AIRES	34.60 S	5.3	4.7	4.1

Table 8.3 The Variation of Insolation with Latitude at 10 Locations. Source: World Bank Group Solar Resource Maps

Notes for above Table 8.3

It must be noted that while the World Bank Group Solar Resource Maps does not provide any data for solar insolation and solar energy potential from latitude 60^0 N and above into the Arctic Region, the US National Renewable Energy Laboratories (NREL) Maps and other Alaska specific reports clearly indicates that Alaska has significant solar insolation and specifically Anchorage, Alaska has an average Solar Insolation of 3.65 Kwh/m²/day and thereby has significant PV potential (PVP).

DNI – Direct Normal Irradiance. This is "the total amount of solar radiation received per unit area by a surface that is always perpendicular to the Sun's rays that come in a straight line from the direction of the Sun at its current position in the sky".

GHI- Global Horizontal Irradiance. This is "the total amount of solar radiation that is received per unit area by a surface that is always positioned in a horizontal manner".

PVP- Photovoltaic Potential. This the energy available to produce electrical energy through the photovoltaic process using the available solar radiation.

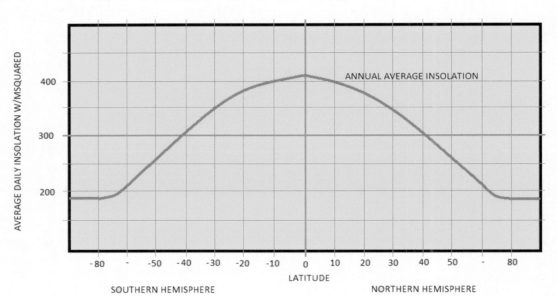

Fig 8.10 Typical Latitude Versus Average Daily Insolation Graph.

Table 8.3 and FIG 8.10 clearly highlights the relationship between latitude and insolation, they also highlight that insolation decreases as the latitude increases northward or southward and also the fact that insolation is highest around the region of the equator covering the tropical region enclosed between 23.5^0 South and 23.5^0 North.

The insolation impinging on the surface of the Earth usually causes an increase in the surface temperature and the highest temperatures typically occurs after the longest periods of insolation with the longest periods of insolation normally occurring during the longest day and the increase in surface temperature of the Earth tend to decrease relative to the decrease in insolation with an increase in latitude, North or South of the Equator.

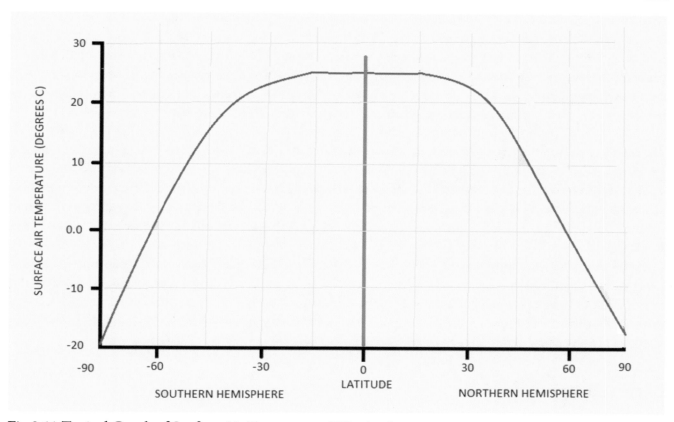

Fig 8.11 Typical Graph of Surface Air Temperature V Latitude.

The graph shown in FIG 8.11 is an approximation of the surface air temperatures in both the southern and northern hemispheres and these temperatures reflects the solar insolation impacting the regions both north and south of equator.

Amount of Insolation Available

The amount energy produced by the Sun that reaches the surface of the Earth is far beyond what the population of the Earth can utilize now or in the future as according to the US Department of Energy(DOE) 430 quintillion (430,000,000,000,000,000,000) joules of energy, approximately 120,000,000,000,000 KW, from the Sun reaches the surface of the Earth each hour of daylight and the population of the Earth uses approximately 410 quintillion joules per year, which indicates quite clearly that the energy supplied by the Sun in one hour is more than enough to meet the energy demand of the Earth for one year. According to Ritchie and Roser (2020) solar energy currently supplies only approximately 1.1% of the current energy demand of the Earth, a clear indication that enough is not being done to utilize more of the natural energy of the Sun. This underutilization of this vast source of free energy may be an indication of two things, the absence of the appropriate technology to adequately reap this energy bounty or a lack of desire on the part of humans, as the data clearly indicates that with the requisite technology the Sun could supply all of the electrical demands of the planet.

The energy from the Sun is spread over the surface of the Earth in amounts that is dependent upon latitude, season, landforms, time of day, seasonal weather conditions and varies from location to location and over the last 40 years solar insolation data has been collected from almost every location on the surface of the earth by a number of institutions inclusive of the US Department of Energy(USDOE), the US National Aerospace and Science Administration (NASA), The US National Renewable Energy Laboratories (NREL) through its National Solar Radiation Database(NSRDB), the US National Oceanic and Atmospheric Administration (NOAA) through the National Centres for Environmental Information(NCEI), SOLARGIS, the World Bank Group and many other international organizations. The collected data has been used to create solar maps that provide readily available information to researchers and professionals in the field of solar energy and this information can be accessed through the different institutions and through publicly available sources such as SOLARGIS, Global Solar Atlas, Worldwide Solar Maps and proprietary sites.

Photon Energy

The other very important component in the production of solar energy comes directly from the Sun, Solar radiation that comes in the most minute packets of energy, these packets are referred to as photons. A photon is a tiny particle, that comprises electromagnetic radiations that has no charge and no resting mass, that travels at the speed of light, has a specific wavelength and frequency and behaves very much like an electromagnetic wave. This photon is also described as the smallest possible packet of electromagnetic energy that has energy ranging from the high energy end of the spectrum, gamma rays and x-rays, down to the low energy end of the spectrum, infrared waves and radio waves, all of which travel at the same speed, the speed of light. These photons are small and are usually described as discrete indivisible energy packets that are stored as oscillating electric fields which may oscillate at varying frequencies and they are normally transmitted over great distances at the speed of light and normally reaches their destination without any significant reduction in speed or energy, except in the case of attenuation caused by natural elements.

These photons while having no mass have been observed to have a momentum which can be utilized to produce a wide spectrum of useful work upon interaction with certain light sensitive materials inclusive of the chlorophyl in plants, transition elements and semiconductors. The interaction with natural or other material produces specific results which are created by the imparting of a photon of energy to the systems with which they interact. The results of photons impacting material usually produce the following:

- The photons give up some or all of its energy to electrons in the material.
- The energized electrons will ultimately transfer the gained energy to its parent material
- High energy photons will penetrate deeper into material than charged particles of similar energy and will dislodge electron from the atoms of the material which it is impacting

Generally, when photons impact the target material one of three things will happen:

- A photoelectric effect will be produced
- The creation of the Compton Scattering Effect
- The production of electron pairs

Photoelectric Effect

During the process of producing the photoelectric effect one of several things will happen inclusive of the photoelectric absorption and the ejection of an electron with high energy with silicon semiconductor and to create an electron and hole pair (an exciton) in an organic material and the movement of electron across boundaries and this shall be as follows:

- In the process of photoelectric absorption, a photon impacts an atom in the target material and completely disappears.
- An energized electron, photoelectron, is ejected from the bound shell of the impacted atom.
- With a photon of high energy gamma rays, the ejected electron will usually be removed from deep in the atom or a K shell as shown in the diagram below.
- The refilling of the inner shell with an electron from the outer shell will produce fluorescent radiation.

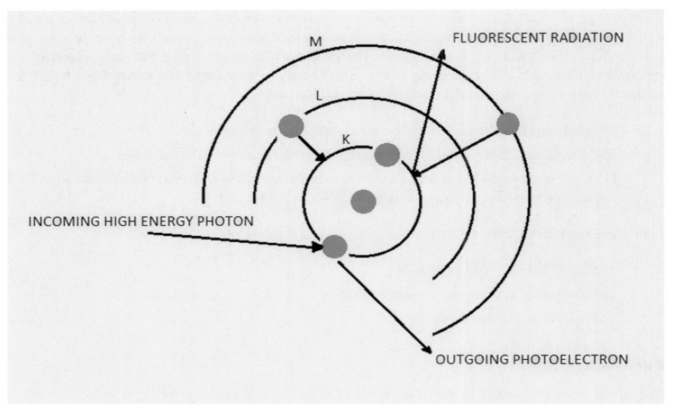

Fig 8.12 Typical Diagram Of High Energy Photon Penetrating The Inner Shell Of An Atom.

The photoelectric effect is usually the major output from photon impact and usually occurs:

- At relatively low photon energy
- With materials with high atomic numbers

COMPTON SCATTERING

Compton Scattering usually takes place after the impact of a photon of gamma radiation and an atom in the target material and it is usually the predominant reaction mechanism for gamma ray photons with targets of metals and tissue.

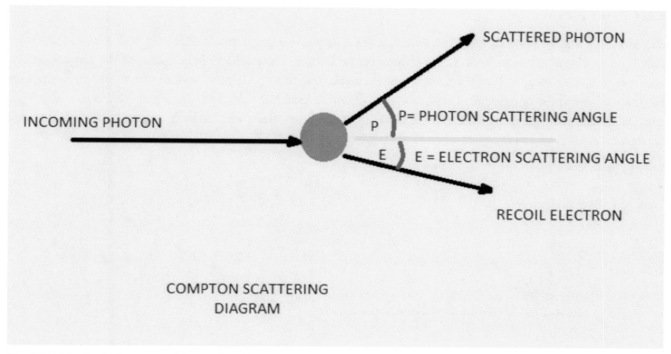

Fig 8.13 Typical Compton Scattering Diagram.

In the Compton scattering three things will happen after the incoming gamma ray photon collides with target material, the incoming gamma ray photon is deflected through an angle P relative to the original direction, the photon will gives up a portion of its energy to the electron ejected from its orbit, the ejected electron is called the recoil or Compton electron, and the ejected photon will be deflected through an angle E relative to the original photon direction, a different direction from the deflected photon. In this Compton Scattering process, the following results are all possible:

- The angles through which the photon and electron can scatter are unlimited.

- The energy given to the electron can vary from a very large portion of the gamma ray energy to almost none of the energy.

- The Compton Scattering is of greatest importance in the field of medicine, for impact with soft tissues, in energy levels ranging from 100 keV to 10 MeV

211

Pair Production

When a high energy photon, higher than 1.022 MeV, impacts a target material it may create a process called Pair Production, which occurs when the photon passes close to the nucleus of the atoms of the target material and encounters a strong field effect from the nucleus subsequent to which the photon disappears and reappears as a pair of electrons, one negatively charged and the other positively charged. These two electrons are not old electrons that were ejected from their orbits but they are two new electrons that were created by the energy mass conversion created by the disappearance of the high energy photon.

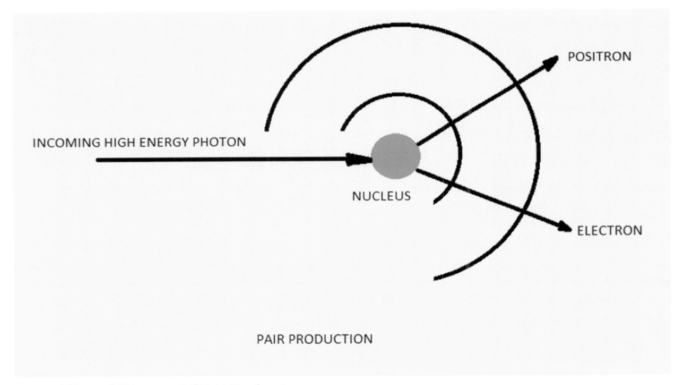

Fig 8.14 Typical Diagram Of Pair Production.

The likelihood of producing an electron pair increases with the increasing energy of the photon and the increasing size of the atomic number of the target material.

In order to produce any of the effects described above, photoelectric effect, Compton Scattering or Pair Production the incoming photon, from any end of the spectrum, must possess a certain level of energy and momentum. According to Plank's equation the energy of any single photon of light is given by the following equation:

$$E = hf_{ew}$$ where h= Planck's constant, f_{ew} = frequency of the electromagnetic

wave and E = Photon Energy. Or E = h.c/wl, where c= wl.f_{ew} = the speed of light, wl = wave length of the electromagnetic wave

In standard physics momentum = mV, where m = to the mass of the body, V = the velocity at which the body is moving, however, because a photon has no mass this equation will not apply and a momentum equation is required for a single photon and is as given in the following equation: photon momentum, pm = h/wl

From the two equations

- E = hfew
- PM = h/wl

A photon will have higher energy with high frequency electromagnetic radiation and a greater momentum with shorter wavelength electromagnetic waves, both of which gives a single photon a greater capacity to eject electrons from their orbits in target material. Electromagnetic waves do not move as single photons but as swathes of photons referred to as "photon flux" which is the number of photons per second in a single beam of sunlight or other electromagnetic radiation and typically one photon of visible light has approximately 10^{-19} joules of energy while the amount of energy in a Photon Flux (PF) given by PF = P/$hf_{ew,}$

Where P = Beam power in watts.

The typical Photon Flux Densities (PF/A) for typical sources of electromagnetic radiation are as shown in the table below.

LIGHT SOURCE (Electromagnetic Radiation)	MEAN PHOTON FLUX DENSITY (photons/s.m^2)
Laser Beam (10 mW, He-Ne, focused to 20 micro meters)	10^{26}
Laser Beam (1 mW, He-Ne)	10^{21}
Bright Sunlight	10^{18}
Indoor Light Level	10^{16}
Twilight	10^{14}
Moonlight	10^{12}
Starlight	10^{10}

TABLE 8.4 Typical Mean Photon Flux Densities for a Sample of Light Sources.

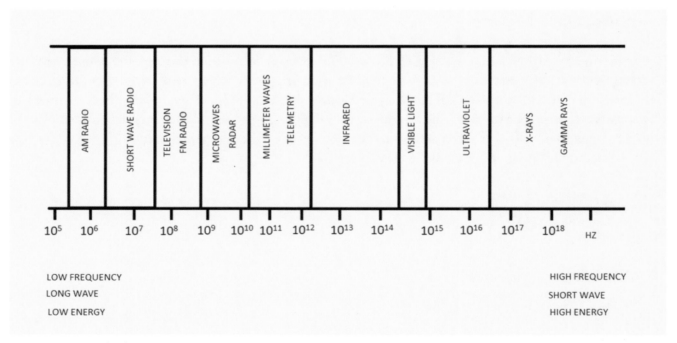

Fig 8.15 Typical Photon Energies For The Electromagnetic Spectrum. Source: Gsu.edu

According to Millikan, the lowest level of energy needed to disconnect an electron from the surface of target material is usually referred to as the photoelectric work function and the threshold energy is a function of the minimum frequency and maximum wavelength that will produce the work function. The minimum frequency is 4.39×10^{14} Hz and the maximum wavelength 683 nm and when either value is plugged into Planck's photon energy equation $E = hf_{cw} = h.c/wl$, gives a minimum photon energy of 1.82eV.

Each target material will have a different work function and some material may have more than one work function depending on the crystalline status of the material and some of these values are highlighted in the table below.

ELEMENT	WORKFUNCTION (eV)	ELEMENT	WORKFUNCTION (eV)
Aluminium	4.08	Magnesium	3.68
Beryllium	5.0	Mercury	4.50
Cadmium	4.07	Nickel	5.01
Calcium	2.90	Niobium	4.30
Carbon	4.81	Potassium	2.30 & 2.29
Cesium	2.10	Platinum	6.35
Cobalt	5.00	Selenium	5.11
Copper	4.70	Silver	4.26-4.73
Gold	5.10	Sodium	2.28 & 2.36
Iron	4.50	Uranium	3.60
Lead	4.14	Zinc	4.30

Table 8.5 Typical Work Function Of Selected Metals. Source: Gsu.edu

CHAPTER 9

Emerging Solar Energy Technology

The utilization of solar radiation to create electrical energy is based on the reaction of some materials to photons of sunlight impinging on their surfaces to dislodge electrons which consequently produce the phenomenon called the photovoltaic effect, a flow of electrons or electricity, due to photoconductance. This phenomenon, has since the creation of the first solar cell focused solar energy research into finding other materials that will behave in a similar manner to produce the required dislodging and the consequent flow of electrons or to finding materials that will produce even greater flows of electrons than the very first material used to demonstrate what became known as and is still called the photovoltaic effect. Solar energy research has also sought to improve the rate at which the Sun's energy can be converted by these materials to electricity and to create the best configuration of solar cells and other materials to most effectively collect and convert the incident solar radiation.

Material

The PV solar energy industry is based completely on the behaviour of photosensitive materials most of which were semiconductors, in the early days of the industry, but these semiconductors have now been complimented by metals, rare metals, alloys, compounds, exotic organic polymers and other new materials at different stages of development. Some of these semiconductors may be described as elemental semiconductors which are pure materials such as antimony, arsenic, boron, carbon, silicon, selenium, germanium, sulphur and tellurium and the other semiconductors are compounds or alloys of other materials such as binary compounds gallium arsenide, lead sulphide and cadmium selenide and ternary compounds such as Mercury Indium Telluride, all manmade material. Semiconductors material do not conduct heat or electricity all the time like metals do and they do not fully block the flow of heat and electricity like non-conducting insulators do, this behaviour places semiconductors directly between metals, conductors, and insulators and this behaviour is a function of the atomic structure of these semiconductors relative to those of metal and non-metals.

All materials are made up of atoms which are the smallest indivisible part of matter and each atom is made of neutrons, protons and electrons, with the neutrons and protons located in the nucleus of the atom while the electrons move around the nucleus. The number of electrons in each atom is a function of the type of material with each material having a different number of neutrons, protons and electrons and each of these particles have different electrical charges, the neutrons are neutral, the protons are positively charged and the electrons are negatively charged. The electrons that move around the nucleus usually have different energy levels depending on their proximity to the nucleus, with the electrons closest to the nucleus having lower energy levels and those further away having much higher energy levels. The different energy level electrons are said to exist in different energy shells with higher energy level electrons existing in the outer shell and the lower energy level electrons existing in the inner shell, with the electrons in the outer shells performing all interactions with other atoms. The usual activities performed by an atom includes forming compounds with other atoms, transferring heat and electricity. The electrons in the outermost shells are usually referred to as valence electrons and the energy shell that they are in is usually referred to as the valence band, which is surrounded by an empty band that is referred to as the conductance band, which is normally empty. The ability for an atom to form compounds, transfer heat and electricity defines the nature of all elements or compounds, with the atoms that quickly transfer heat and electricity being described as conductors, the atoms that do not transfer heat or electricity are described as insulators and the atoms that are not good conductors or good insulators and only transfer heat and electricity under certain conditions but not under others are usually described as semiconductors.

The ability to perform the functions mentioned are dependent upon the ability of the electrons to move between the valence band and the conductance band and this ability to move between the bands is a function of the distance between the bands, with conductors having valence and conductance bands that overlap, insulators having bands that are far apart and semiconductors having bands that are fairly close together. In conductors where the two bands are overlapping very little energy is required to move an electron from the valence band to the conductance band while semiconductors with a small band gap will require a comparatively larger amount of energy to bridge the energy band gap to facilitates the movement of electrons from the valence band to the conductance band. With insulators having a larger energy band gap the amount of energy required to bridge this gap would fairly high, in high enough to consume the insulator before any electron transfer could take place, which make the transfer of electrons in insulators very unlikely even with the application of large amounts of heat or electricity. The energy band gaps of conductors, insulators are highlighted in the Fig 9.1 below.

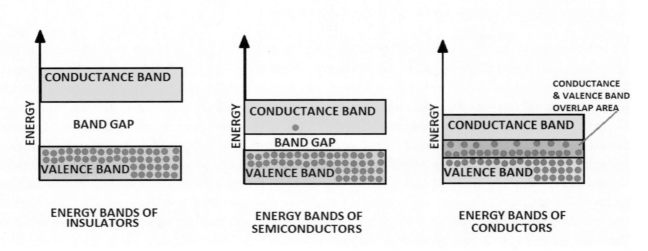

Fig 9.1 Typical Diagram Highlighting Band Gaps for Conductors, Semiconductors and Insulators

The earliest materials that clearly demonstrated the photovoltaic effect were semiconductors with Selenium being the first, followed later by the Silicon, Germanium, Tin and Tellurium and much later by alloys and compounds of Copper, Gallium, Indium, Cadmium, Germanium, Cadmium and Lead. Silicon is currently used to make more than 90 percent of all semiconductors used in the solar energy industry, however, silicon like most semiconductor material does not occur in its desired form and must be mined then processed to produce the desired form, 99.9999 percent pure silicon after which the silicon must be shaped and formed into wafers that are six inches in diameters and two hundred to three hundred nano-meters thick. The huge amount of processing required and the large volume of silicon material that must be used for each solar cell makes silicon cells very expensive to produce and this relatively high cost has fostered the desire for the development of the next generation of solar cells, even though silicon solar cells had proven to be relatively efficient.

The second-generation of solar cells were made by utilizing a different technology and different materials inclusive of the binary and ternary compounds and the new technologies fostered the development of systems that achieved the goal of reducing the overall cost to produce and utilized solar energy. The technology utilized was called thin film technology and it utilized very thin films of semiconducting material to produce solar cell and consequently electrical energy for public consumption, however, while meeting the goal of utilizing less costly material that would lower the overall cost to produce solar cells, the new solar cells did not exceed the performance of the older silicon cells, but had a far lower conversion efficiency. This technological shortfall ensured that efforts continued to find and develop more efficient alternative solar technologies that will eventually replace the current leader of the pack, the expensive silicon technology and bring up the current generation of solar cell, the third generation.

The research work to produce the third generation of solar cell technologies is ongoing with many new technologies being reviewed, sampled, produced for trials and tested in laboratories all around the world and some of these new technologies includes Perovskite solar cells, organic solar cells, quantum dot solar cells and hybrids of all of these new technologies. The results produced to date has seen the development and production of solar cells that are being fabricated from newer material inclusive of organic polymers that requires even less material than the second-generation of thin film solar cells, with very high hopes of greater improvements in conversion efficiency as each new technology matures to the point of making them suitable to be placed on the market. All of the improvements in solar cell technology created to date were as a result of great work done in the field of material science, material science research that was at first completely focussed on natural semiconductors but later transitioned to metals, compounds of these metals and alloys of these metals that are capable of producing the photovoltaic effect and now continues with the focus on organic polymers which have the potential to take the solar energy industry into uncharted waters.

Fundamental Material Requirements

All materials that have been utilized in the development and fabrication of solar cells have one property in common, they are all sensitive to light, photosensitive, the first requirement to make them suitable for the production of electrical energy from sunlight.

Photosensitive Materials

Photosensitive materials have several very important properties inclusive of the following, they are sensitive to light energy, they have the ability to absorb light energy and the capacity to have its behaviour modified by the impacted of the light energy on its exposed surface. The basic properties responsible for the behavioural changes that occur in the photosensitive materials include the ability of the material to quickly absorb a photon of light, the ability to utilize a portion of the energy of the photon that was absorbed to eject an electron from its valence band into the conductance band, the ability to transfer the ejected electron across boundaries and terminals to create and establish a flow of electrons and the ability to temporarily change their electron configuration when they lose an electron or undergo changes in either their molecular or interfacial properties or both. With any or all of these changes these materials may adjust their natural behaviour to become more active in one or more ways inclusive of becoming photoconductive, becoming photoluminescent, becoming a photocatalyst and demonstrating a photovoltaic effect (photon energy conversion). These demonstrated behavioural changes caused by their reaction to absorbing, storing and converting light energy gives these materials the ability to act as solar cells, photodiodes, photodetectors, photographic material and photocatalysts.

If the interactions of these materials with solar radiation is sufficiently strong enough these materials may also act upon their molecular environment and create changes to their chemical structures to form energy-rich compounds and some photosensitive material may also be "doped" with other material to facilitate the desired output or to amplify their natural output that occurs after interacting with the light energy.

Photoluminescent Material

According to Gfroerer(2006) one definition of photoluminescence is that it "is the spontaneous emission of light from a material under optical excitation" and this optical excitation occurs when the material is impacted by a photon of light which dislodges an exciton, electron-hole pair, utilizing only a portion of the energy supplied by the photon and sending the excess energy to the material, which the material then dissipates, the excess energy, in one of several ways inclusive of giving off light, luminescence, which is labelled photoluminescence as the event was initiated by the action of a photon of light. The energy of the luminescence is the difference between the photon energy and the energy required to eject the exciton. There are three modes of photoluminescence and they are as follows:

- **Resonant Radiation**. In resonant radiation the initial photon of light, at a specific wavelength, impacts the material and a secondary photon of light at the same wavelength is emitted without any significant change in energy and this event takes place in a very short period of time, 10 nanoseconds. This resonant radiation is considered to be another type of fluorescence.

- **Fluorescence**. Fluorescence occurs when an atom in a material absorbs the energy of a photon of light, UV or Xray, which raises the energy level of the atom beyond its ground state and when this atom is cooled and returns to its ground state it gives off the excess energy in the form of light. The amount of light given off and the frequency of occurrence is always a function of the material and the energy contained in the light source.

- **Phosphorescence**. Phosphorescence occurs when a material is impacted by radiation, UV -IR, Xray or Gamma Ray, and this impact causes the excitation of the electrons in the material raising the electrons to new energy levels. When the electron returns to its ground state energy level it emits the excess energy as visible light which persist after the radiation source was removed, for duration varying from fraction of seconds to hours.

A small group of material demonstrate resonant radiation, fluorescence and phosphorescence and this group include phosphors and the wide band-gap group of semiconductors inclusive of Silicon Carbide (SiC) and Gallium Nitride (GaN).

Photocatalyst

The most basic definition of a photocatalyst is that it is a material that absorbs photons of light which raises its energy above its ground state energy and provides this increased energy to initiate a chemical reaction in other substances. This capacity is utilized to facilitate the decomposition of harmful substance inclusive of, stains of oil and fats, organic stains, obnoxious smell, nitrogen oxides (NO_x), antibacterial and other harmful substances. The capacity to perform these functions makes photocatalysts and the process of photocatalysis highly valuable for use in energy production, wastewater treatment and air purification. The main photocatalyst currently in use is titanium dioxide (TiO_2) and the polymorphs of titanium dioxide (TiO_2), with anatase showing the greatest catalytic effect. The basic mechanisms in such operations occurs as follows:

- Photons of UV light are utilized to generate pairs of electrons and holes, excitons, from titanium dioxide (TiO_2).
- The generated electrons and holes are then utilized to create the necessary reactants. The electron reacts with O_2 to produce the O_2^- anion radicals and the hole reacting with water to produce active OH radicals.
- The two products, O_2^- anion and active OH radicals, utilize their oxidizing and reducing capacities to decompose the objectionable stains, obnoxious smells, NO_x and other substances.

Photoconductive Material

Photoconductivity speaks to the impact that light has upon the electrical conductivity of some solid materials and this phenomenon is utilized in many areas inclusive of the ship and aircraft detection systems, photovoltaics, television cameras, light meters, infrared detectors, voltage regulators, relays, and other electronic controls. According to Qiu et al (2021) "Photoconductivity is an optical and electrical phenomenon in which a material becomes electrically conductive owing to the absorption of electromagnetic radiation such as visible light, ultraviolet light, infrared light or gamma radiation." Basically, stating that the resistance of these materials to the passage of an electric current can be altered due to the presence of light, with additional explanation as follows:

- The absorption of electromagnetic radiation by a material causes the number of free electrons and holes (excitons) to increase and this cause the electrical conductivity of the material to increase.

- To be effective in raising the excitation levels of the electron and hole pair, the light that strikes the material must have a certain level of energy that is sufficient to liberate the electrons and thereby creating the flow of electricity.

There are basically two classes of photoconductive material idiochromatic and allochromatic materials.

Idiochromatic Material

Idiochromatic materials are materials that show photoconductivity in their pure state and this photoconductivity is directly related to the intrinsic properties of this material alone and some of the typical idiochromatic materials are cadmium, copper, lead, iron, manganese, nickel, chromium, titanium, vanadium and compounds of these elements inclusive of lead sulfide, lead selenide, cadmium sulfide and mercury cadmium telluride. All of these materials are known as transition elements and the compounds of these transition elements. This group of elements consist of 68 elements divided into 40 transition elements and 28 inner transition elements and the basic physical properties of these transition elements are as follows:

- They are mostly metals with high melting and boiling points with the exception of mercury which is a liquid at room temperature.

- Low energy electron transitions facilitated by infrared, visible and ultraviolet radiations have been observed for the free element, their compounds and complexes, that is they have fairly weak bonds in their outer electron orbit or 3rd level orbit and lose electrons more easily than alkali metals.

- They are lustrous and malleable and can be hammered or bent into shape easily

- They are very good conductors of heat and electricity

- They are mostly hard and tough, with mercury being that exception as a liquid

- They have high densities

- Many of the group are paramagnetic and can be used to construct paramagnetic compounds

Along with these physical properties the transition elements have the following chemical properties:

- As a group, transition elements are found to be less reactive than the alkali metal group inclusive of lithium, sodium, potassium, cesium and francium.

- They tend to form colored ions with different charges

- Some members of this group such as platinum, silver and gold tend to be highly unreactive

- Many members of this group have the capacity to be used as catalysts.

- They have variable oxidation states

- They create stable complex compounds

- They have a high charge/radius ratio

- The have variable oxidation stages

- They could be used to create active catalytic compounds.

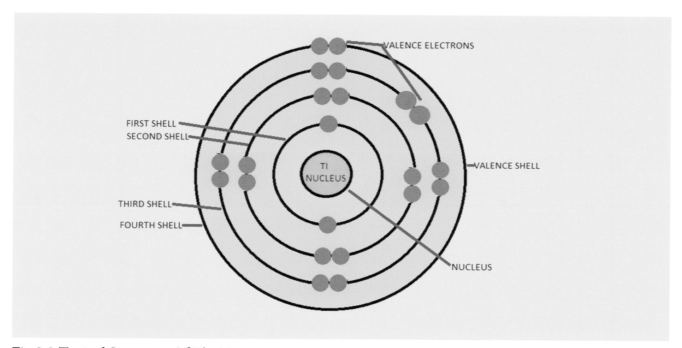

Fig 9.2 Typical Structure Of The Titanium Atom A Transition Metal.

As previously discussed, all elements are made up of atoms and these atoms determine the characteristics of each element, their structure, their physical status as liquid, solid or gas, their interaction with other elements and their physical characteristics. The atoms of each element have a nucleus consisting of protons and neutrons surrounded by several shells of electrons and the electrons, which are negatively charged particles, are responsible for the interaction that elements may have with each other. These electrons are usually located in several energy shells around the nucleus and each of these energy shell can hold varying number of electrons with the first shell capable of holding two electrons, the second shell capable of holding up to eight electrons, the third shell capable of holding up to 18 electrons and the fourth shell capable of holding up to 32 electrons. The electrons in the outer shells usually have the highest energy levels and usually require less energy than the electrons on the inner shells to dislocate and transfer and it is usually these electrons which are given up to forge bonds with other elements in the process of forming or creating compounds. The electrons in the outer shell are usually referred to as the valence electrons and most transition metals usually have two electrons available for forming chemical bonds with other elements, however, as shown in the diagram in FIG 12.1 above, these transition elements may also give up electrons in the next closest shell also to form necessary bonds.

The number of valence electrons typically available for transition elements are two and one, with most having two valence electrons and the exceptions copper and chromium having one valence electron each, however, with their ability to have multiple oxidations states these same elements may have seven, six, five, four, three, two and one valency electrons depending on the compounds that they form. The table below highlights the varying oxidation states for the first row of the transition metals in the Periodic Table.

Element	Atomic Number	+ 1	+2	+3	+4	+5	+6	+7
Scandium	21	No	Rare	Common	No	No	No	No
Titanium	22	No	Rare	Rare	Common	No	No	No
Vanadium	23	Rare	Common	Common	Common	Common	No	No
Chromium	24	Rare	Common	Common	Rare	Rare	Common	No
Manganese	25	Rare	Common	Common	Common	Rare	Common	Common
Iron	26	Rare	Common	Common	Rare	Rare	Rare	No
Cobalt	27	Rare	Common	Common	Rare	Rare	Rare	No
Nickel	28	Rare	Common	Rare	Rare	No	No	No
Copper	29	Rare	Common	No	No	No	No	No
Zinc	30	No	Common	No	No	No	No	No

Table 9.1 Typical Data on Oxidation States of Row 1 of the Transition Metals.

Ionization and Bonding Energy for Transition Metals

Element	Atomic No./ and Symbol	1St Ionization Energy	2nd Ionization Energy	3rd Ionization Energy	4th Ionization Energy
Scandium	21/ Sc	633	1,235	2,389	7,091
Titanium	22/ Ti	659	1,310	2,653	4,175
Vanadium	23/ V	651	1,414	2,830	4,507
Chromium	24/Cr	653	1,591	2,987	4,743
Manganese	25/Mn	717	1,509	3,248	4,940
Iron	26/Fe	763	1,562	2,957	5,290
Cobalt	27/Co	760	1,648	3,232	4,950
Nickel	28/Ni	737	1,753	3,395	5,300
Copper	29/Cu	746	1,958	3,555	5,536
Zinc	30/Zn	906	1,733	3,833	5,731

Table 9.2 Typical Ionization Energy Data for Some Transition Metals. Source: Depauw.edu

ELEMENT	ATOMIC No./ SYMBOL	1ST Ionization Energy kJ/mol	2nd Ionization Energy kJ/mol	3rd Ionization Energy kJ/mol	4th Ionization Energy kJ/mol
Silicon	14/Si	787	1,577	3,232	4,356
Gallium	31/Ga	579	1,979	2,963	6,180
Germanium	32/Ge	762	1,538	3,302	4,411
Arsenic	33/As	947	1,798	2,735	4,837
Selenium	34/Se	941	2,045	2,974	4,144
Silver	47/Ag	731	2,070	3,361	-
Cadmium	48/Cd	868	1,631	3,616	-
Indium	49/In	558	1,821	2,704	5,210
Tin	51/Sb	709	1,412	2,943	3,930
Tellurium	52/ Te	869	1,790	2,698	3,610

Table 9.3 Typical Ionization Energy Data for Material Already Used as Semiconductors for Solar PV Systems. Source: Depauw.edu

Allochromatic Material

Allochromatic materials are materials that display photoconductivity when associated with other materials or an impurity or an imperfection in the lattice structure of the material. Typical allochromatic material include germanium and silicon as hosts and transition metals such as cobalt, nickel, vanadium, titanium, arsenic, copper, gold and indium as impurities.

Persistent Photoconductivity

As previously defined by Qiu et al (2021) the impact of electromagnetic radiation inclusive of visible light, ultraviolet light, infrared light and gamma radiation, on particular types of material causes a change in the electrical conductivity of the material and this change in electrical conductivity due to the impact of light is normally referred to as photoconductivity. This induced change in conductivity is usually temporary and only occurs in the presence of the light source, however, in some cases even with the removal of the light source the target material continues to behave as if the light source is still present and this effect can continue for extended periods of time depending on the specific material and this phenomenon is usually referred to as persistent photoconductivity (PPC). In some cases, the photoconductive effect is not only persistent but is also very large and may last anywhere from a few hours to days depending on the material and surface modification of the material and is referred to as large and persistent photoconductivity (LPPC). According to Yin et al (2018) LPPC in semiconductors is caused by the trapping of photogenerated minority carriers at crystal defects and the phenomenon, LPPC, has been confirmed to occur in several semiconductor material inclusive of silicon, group III -Vs semiconductors, oxides and chalcogenides and specifically it has been reported that cadmium sulphide (CdS) has also demonstrated this phenomenon, the same cadmium sulphide that is widely used in thin film solar cells. The result of recent experimental work has also shown that the use of chemical baths depositions can be used to enhance the photoconductivity of cadmium sulphide thin film to over 9 orders of magnitude above standard cadmium sulphide thin film cells and improve photoconductivity decay time from mere seconds to 10^4 seconds.

New Material

Historically, the photoconductive effect was typically observed in traditional semiconductors and alloys of metals inclusive of selenium, silicon, copper, indium, gallium, germanium, zinc, cadmium, magnesium, sodium, potassium, cesium, rubidium and alloys of these metals, however, since the 1960s research work has shown that organic elements have demonstrated the ability to show photoconductive effect and even persistent photoconductance, PPC, and large persistent photoconductance, particularly in organic polymers.

Photoconductance in Organic Polymers

A breakdown of the word polymer will reveal its most basic meaning, things of many parts, poly meaning many and meros meaning parts, from the Greek root words and according to Helmenstine (2019) "A polymer is a large molecule made up of chains or rings of linked repeating subunits which are called monomers" and this large molecule can consist of hundreds or thousands of atoms all joined together in a unidimensional arrangement. It has also been established that there are basically two types of polymers, organic and inorganic polymers, which can be further classified into natural biological polymers and synthetic, manmade, polymers. The organic biological polymers or organic biopolymers consist of silk, natural rubber, cellulose, wool, amber, keratin, collagen, starch, DNA, and shellac. While the manmade organic polymers consist of polyvinylchloride, polystyrene, synthetic rubber, silicone, polyethylene, neoprene and nylon. In general, polymers are made up of a backbone or skeletal framework of one element with one or two side elements and organic polymers will have a carbon skeletal framework while inorganic polymers will not have a carbon skeletal framework. The side elements for organic polymers may be small and include such elements as chlorine or fluorine, or hydrogen, or it may be long chains of alkyl units or long aryl units.

Organic Biological Polymers

As stated above when monomers join together, they form giant molecules or polymers which are also called macromolecule and these natural macromolecules are used to build tissue and other components in living organisms. While there are many macromolecules, all of these are made up from only 50 monomers which combine in many different sequences to create the many unique macromolecules that make up all living things in the many and varied species of different shapes, sizes and colors that exist on the planet. In general, there are four types of macromolecules biopolymers inclusive of carbohydrates, lipids, proteins and nucleic acids and these polymers consists of different monomers which all serve different functions.

Synthetic Organic Polymers

Synthetic organic polymers are manmade polymers that were created from petroleum oil and include products such as nylon, synthetic rubber, polyester, Teflon, polyethylene and epoxy that have found wide use in industry, manufacturing and household products. A few of the products made from these polymers include bottles, pipes, plastic containers, insulation for electrical wires and conductors, clothing, toys and non-stick cooking utensils.

Polymers in Solar Energy

Organic polymers are now becoming very important in the field of solar energy as extensive research has been ongoing to ascertain their photoconductive properties, their ability to produce electrical energy efficiently, their durability and their ability to be mass produced as solar cells using polymers at a cost-effective rate. Work to utilize other materials inclusive of organic polymers in solar energy systems has been ongoing from the 1960s and the most significant results produced since that time have been the following systems Dye Sensitized Solar Cells, Perovskite Solar Cells, Organic Cells, Organic Tandem Cells and Quantum Dot Cells, all of which are based on thin and thick film technology.

Dye Sensitized Solar Cell

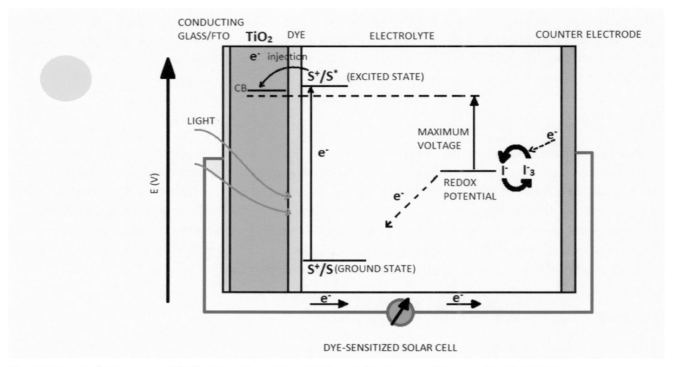

Fig 9.3 Typical Diagram Of The Dye-Sensitized Solar Cell. Source: Sharma Et Al (2018)

The main components of the dye-sensitized solar cell (DSSC) are the working electrode, the sensitizer (dye), the electrolyte and the counter electrode and these four components are assembled as follows:

- The working electrode/photoanode. This is usually constructed of a transparent conducting oxide (TCO) glass, which is a glass substrate that has been coated with fluorine-doped tin oxide (FTO) or indium-doped tin oxide (ITO) after which a layer of titanium dioxide is deposited on it by one of three methods, doctor blading, screen printing or inkjet printing. This layer of titanium dioxide is a mesoporous electron transferring layer that is essential to the movement of electrons across the cell.

- Dye/Sensitizer. This is a single layer of dye that is spread on and bonded to the porous titanium dioxide which acts to increase the sensitivity of the working electrode, allowing it to absorb more photon energy.

- Liquid Electrolyte. This electrolyte contains several very important elements which are vital to the operation of the cell and these elements include iodide/triiodide, an organic solvent and other essential additives.

- Counter Electrode (CE)/ Cathode. This element is similar to the photoanode as it consists of a transparent conductive glass substrate to which a catalyst, platinum or carbon, has been added. This cathode performs two basic functions, accepting electrons from an external circuit and converting the developed triiodide ions back to iodide ions by the way of a charge transfer process.

A general description of the four components are as follows:

- Transparent and Conductive Substrate. Transparent and conductive substrate is the conductive glass which must have a minimum transparency of 80% to facilitate the passage of the optimum levels of sunlight to reach the working electrode. This substrate must also have a high electrical conductivity and this high level of conductivity is usually achieved by the application of fluorine-doped tin oxide (FTO) and or Indium-doped tin oxide (ITO) to the soda lime glass substrate. The FTO film has a transmission of approximately 75% and a sheet resistance of 8.5 ohms per centi-meter squared and the ITO film has a transmission greater than 80 percent and sheet resistance of 18 ohms per centi-meter squared.

- Working Electrode/Anode. The working electrode/anode is made up of a thin layer of a semiconductive material such as titanium dioxide or Zinc Oxide or Niobium Pentoxide or Tin Dioxide (n-type) and Nickel Oxide (p-type) deposited on the transparent glass plate made with FTO or ITO. Due to its low-cost, ready availability and its non-toxic nature titanium dioxide is usually the preferred semiconductor used, however, to improved it transmission and conductance it is usually immersed in the dye sensitizer.

- Photosensitizer or Dye. The dye or photosensitizer is a most critical element as it is responsible for increasing the capacity of the cell to ensure maximum absorption of the available sunlight and in order to perform this function all dye must have certain basic properties inclusive of

- Must be luminescent

- The spectral range to be covered must include the UV – Visible light region and the near Infrared region.

- The highest energy orbital with one or two electron or the highest occupied molecular orbital (HOMO) should be located far from the surface of the titanium dioxide and the lowest energy orbitals with no electrons or the lowest unoccupied molecular orbital (LUMO) should be placed very close to the surface of the titanium dioxide as possible and therefore should have a higher potential than that of the titanium dioxide conduction band.

- The potential of the HOMO must be lower than that of the redox electrolytes.

- The dye containment must be water tight, resistant to abrasion and leakage to ensure the longevity of the cell and to prevent direct contact between the electrolyte and the photoanode. Any water induced distortion of the dye from the titanium dioxide surface may reduce the stability of the cell.

- The dye cannot work alone in relation to its adherence to the titanium dioxide and other chemicals must be placed between the it and the titanium dioxide and these chemicals include chenodeoxycholic acid, a co-absorbent, alkoxy-silyl, an anchoring element, phosphoric acid and the carboxylic acid group. The main function of these chemicals is to prevent the aggregation of the dye on the surface of the titanium dioxide and to prevent the premature failure of the cell due to the recombination reaction between redox electrolyte and electrons in the titanium dioxide nanolayer. This action also acts to form stabilize linkages that also ensures the longevity of the cell.

- Electrolyte. The electrolyte generally has five main components that include a redox couple, a solvent, additives, ionic liquids and cations and these electrolytes usually include several variations I^-/I_3^-, $Br/Br2$, SCN/SCN_2, $Co(II)/Co(III)$. The main property that electrolyte must possess are as follows:

  - The redox couple must be able to regenerate the oxidized dye in an efficient manner.

  - Must have long-term chemical, thermal and electrothermal stability.

  - Must be non-corrosive relative to the DSSC components

  - Must facilitate fast diffusion of charge carriers, enhance conductivity and create effective contact between the working electrode/anode and the counter electrode/cathode.

  - The absorption spectra of electrolyte must not overlap with the absorption spectra of the dye.

- Counter Electrode/Cathode. The main component of the counter electrode is platinum or carbon and the main function of the counter electrode is the to act as a catalyst in the reduction of the liquid electrolyte and to collect holes from the hole transport material. The counter electrode works in close relationship as they are sealed together with only the electrolyte between them.

General Operation of the DSSC.

The operation of the dye sensitized solar cells consist of four basic steps, light absorption, electron injection, transportation of carrier and collection of current.

Light Absorption. The incident photons of light pass through the glass and the titanium dioxide and is absorbed by the photosensitizer and electrons from the sensitizer absorb some of the energy of the photon energy and gets elevated from its ground state, S^+/S, to an excited state, S^+/S^*, as shown in FIG 12.5 of the dye which corresponds to a photon energy level of 1.72 eV.

Electron Injection. The excited electrons, with a lifetime of nanosecond duration, are injected into the conduction band of the nano-porous titanium dioxide electrode which is located below the excited state of the dye, where the titanium dioxide absorbs a small portion of the photon energy from the Ultra Violet light region of the photon, which results in the oxidizing of the dye.

Transportation of Carrier. The electrons that were injected into the conduction band are transported between nanoparticles and diffuse towards the back contact and travel through the external circuits to reach the counter electrode.

Collection of current. At the counter electrode the transported electrons go to work reducing the I_3 to I^- and the regeneration of the ground state of the dye takes place by the acceptance of the electrons from I^- ion redox mediator and I^- gets oxidized to I_3^-.

Perovskite Solar Cell

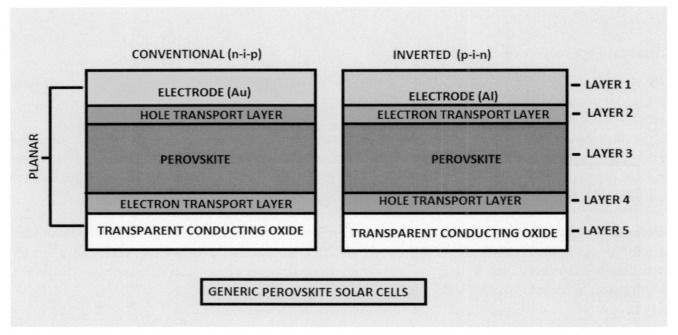

Fig 9.4 Typical Diagram Of Generic Perovskite Solar Cells. Source: D.s. New Energy/Ossila.com

The different layers highlighted on the diagrams contain the following:

- Layer 1: Metal Backed Contact which may be either platinum, gold or silver

- Layer 2 Conventional: Hole transport layer made up of Organic HTMs: SPIRO-MeOTAD, or PTAA and so on. This layer may also be made up of Inorganic HTMs: NiO or CuI or CuSCN and so on.

- Layer 2 Inverted: Electron transport layer made up of Titanium dioxide, Zinc Oxide or Zirconium Dioxide or similar material.

- Layer 3: The active layer which contains the Perovskite which could be one of several formulations inclusive of $CH_3NH_3(PbSnCu)(I,Br,Cl)$ or other formulations.

- Layer 4 Conventional: Electron transport layer made up of Titanium Dioxide, Zinc Oxide or Zirconium Dioxide or similar material

- Layer 4 Inverted: Hole Transport Layer made up of Organic HTMs: Spiro-MeOTAD, or PTAA and so on. This layer may also be made up of Inorganic HTMs: NiO, CuI, or CuSCN or similar material.

- Layer 5: Transparent Conducting Oxide which is glass treated with fluorine-doped tin oxide (FTO) or Indium-doped tin oxide (ITO)

A third version of this Perovskite Solar Cell is shown in the figure below.

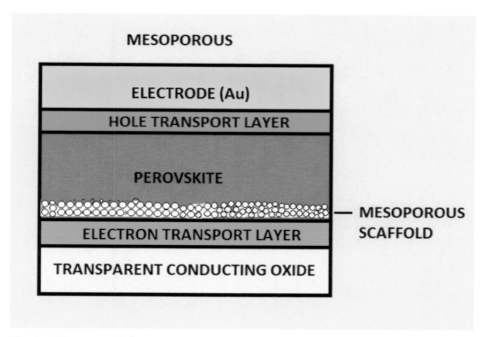

Fig 9.5 Diagram Of Mesoporous Perovskite Solar Cell. Source: D.s. New Energy/Ossila.com

For the mesoporous Perovskite solar cell, the main difference is the addition of the mesoporous scaffold in the third layer, this layer provides a level of control which is essential for the perovskite cell as these cells control crystallization of the perovskite material and charge carrier transport.

Perovskite is a mineral that was first found in the Ural Mountains and it is one of the most interesting new finds in the field of material science that appears to have enormous potential in the field of solar energy and associated electronic fields. The original Perovskite mineral consisted of calcium, titanium and oxygen in the form of the compound $CaTiO_3$ and in this state it forms a particular crystalline structure which may be described as a large atomic structure which is a positively charged ion (cation) of a particular material, type A, at the center of a crystal cube, the corners of the cube occupied by another material, type B, and the faces of the crystal cube are occupied by a third material, type X, with a smaller atomic structure. The typical structure of this crystal is as shown in Fig 9.6 below.

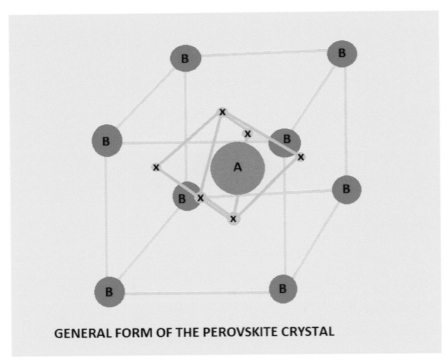

GENERAL FORM OF THE PEROVSKITE CRYSTAL

Fig 9.6 Typical Diagram For The Perovskite Crystal. Source: D.s. New Energy/Ossila.com

This combination of materials and crystal structure gives the perovskite certain characteristics and properties which makes it quite useful in the solar energy and associated electronic industries, some of these properties include superconductivity, giant magnetoresistance, spin-dependent transport and catalytic properties. The desirability of these properties makes Perovskites very valuable and based on these known values the Perovskites structure can be and has been replicated to produce many more crystals and compounds with the Perovskite structure, the basic ABX_3 formulation and crystallographic structure. Most current Perovskites contain the following in the ABX_3 formulation:

- A- A large organic cation – usually methylammonium ($CH_3NH_3^+$) or formamidinium ($NH_2CHNH_2^+$)

- B – A big inorganic cation – usually lead (II) (Pb^{2+})
- X_3 – A smaller halogen anion -usually chloride (Cl^-) or Iodide (I^-)

This standard formulation allows for the creation of numerous perovskites using different materials that match the requirements given above, a large organic cation, a big inorganic cation and a smaller halogen anion, however, the more efficient perovskites have been found to be those that are based on Group IV metal halides (specifically lead).

The Importance of Perovskites

Perovskites while relatively new have to date demonstrated remarkable qualities that make them quite attractive to the solar energy world, where they have been installed and tested since 2012. The results from these and other trials have shown that perovskites compare very well with the existing cell materials inclusive of silicon, the existing thin film material and are mostly much better than other experimental materials. Perovskites compares very well in the following areas:

- Power Conversion Efficiency
- The rapid rate at which improvements are being achieved with the Perovskite technology.
- Perovskite solar cells have in a four years period equalled the efficiency of cadmium telluride (CdTe) and as of June 2018 they have also exceeded all other thin film non-concentrator technologies inclusive of Cadmium Telluride and Copper Indium Gallium Selenide.
- Better conversion of photon energy compared to existing technologies. Perovskites utilize up to 70% of incident photon energy while most other systems only utilize about 50%.
- Costs associated with the high achievements of perovskites have been to date significantly lower than existing state-of-the-art technologies inclusive of Gallium Arsenide (GaAs). They are also close to reaching and by-passing the low cost achieved by the crystalline silicon technologies, which cost 1000 time less than the existing state-of-the-art technologies.

Along with the many positives achieved by the perovskite technologies there are also many negative impacts associated with them inclusive of:

- Environment Impact. Significant negative impact upon the environment due to use of lead as a major component.

- Long-term Instability. Long-term instability is the biggest negative issue associated with perovskites as they are sensitive to water, light, oxygen and heat, due to the basic properties of the perovskites and these issues make them less than suitable for use in solar cells which are all installed and operated in the outdoor environment where there will be copious amount of water, sunlight, oxygen and heat.

- Current-Voltage Hysteresis. This problem is commonly associated with mobile ion migration in combination with high levels of recombination.

Organic Photovoltaics (OPV)

Organic Photovoltaics (OPV) is completely based upon the use of organic semiconducting polymer material for the fabrication of the solar radiation collection and conversion portions of the PV cells. The definition of the organic solar cell, according the Mattoni (2015) is that it is a multilayer device made up of organic compounds with the ability to convert photon energy into electrical energy. The active components of these devices include three basic components, an active layer and two electrodes, an anode and a cathode, which enclose the active layer between them, and a substrate which forms the base upon which the cell is built. Of great importance also, is the type of junctions formed by these three components, especially the junction formed in the active layer.

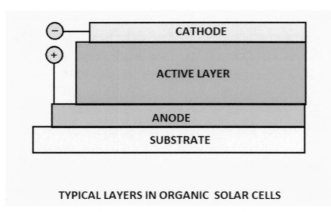

Fig. 9.7 Typical Layers In Organic Solar Cells. Source: Mattoni (2015)

Active Layer

The active layer which is normally $50 - 100$ nm thick consists of two layers of light absorbing organic semiconductors, one donating and the other accepting electrons, that work together to produce the work of the solar cell and these two layers are usually called the donor and acceptor layers as shown in the figure below. The donor layer usually contains organic molecules with low electron affinity and low ionization potential while the acceptor layer usually contains organic molecules with higher electron affinity and higher ionization potential. These two layers are usually made up of several different organic semiconductors, conjugate organic polymers, inclusive of poly-3-hexylthiophene (P3HT), polyphenyl vinylene (PPV), Cyano PPV, thalociline or pentacene as donors and phenyl-C_{61}-buteryc acid methyl ester (PCBM) materials and Fullerene C60 0r C70 as acceptors. The relationship between donors and acceptors must follow a consistent pattern if the transfer of electron is to take place, the donors must easily give away electrons and the acceptor must easily receive the electrons and the electron affinity of the acceptor must be significantly greater than the electron affinity of the donors.

Electrodes

Each organic solar cell has two electrodes one positive (anode) and one negative (cathode) and these two electrodes serve as the points from which the electrical energy that has been created from the photons of light may be extracted from the cell. One of these electrodes is normally made from a metal, platinum, gold, silver, aluminium or some other metals of high conductance, while the other electrode is may be a transparent glass that may be infused or covered with a conductive element such as indium tin oxide (ITO).

Junctions

Typically, there are two basic types of junctions utilized in the organic solar cell active layers, Bilayer Heterojunction and Bulk Heterojunction, as shown in FIG 9.8 below.

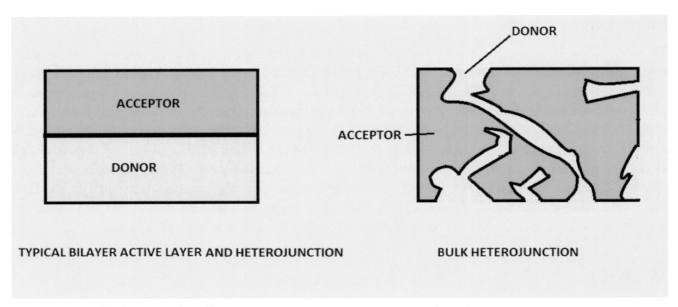

Fig 9.8 Typical Bilayer And Bulk Heterojunction. Source: Mattoni (2015)

The heterojunction is usually the point at which two different semiconductors meet, from which the electron transfer will take place and there are typically two such junctions, the bilayer heterojunction and the bulk heterojunction. In the typical bilayer the two surfaces at the interface tend to be two flat surfaces that meet at the interface, while in the bulk there is a meshing of the two conductors with fingers of each semiconductor interlocking with the fingers of the other exposing each to more surface area than would have been the case with a bilayer and this greater exposure enables greater transfer of electrons, consequently a greater flow of generated electricity.

The basic steps in the operation the organic solar cell are absorption, generation, transportation and collection of the free carriers, as explained below:

- **Photon absorption**. The organic solar cell absorbs the incident photon in the donor side of the active layer after passing through the transparent conductive glass substrate.
- **Exciton Formation**. The absorbed photon dislodges an exciton, a combination of a strongly bound electron- hole pair, from the donor material.
- **Exciton Diffusion**. The exciton is diffused, transported, to the donor-acceptor interface.

- **Exciton Dissociation**. The exciton is separated, dissociated, by charge transfer at the donor-acceptor interface into a free electron and a hole.
- **Electron Hole Pair Formation**. The electron and hole are separated from the interface and move in different directions, the electron into the acceptor leaving the hole behind with the donor.
- **Charge Transfer and Collection**. After the separation of the electron and hole pair at the interface, the electron moves in the direction of the cathode while the hole moves in the direction of the anode and the flow of electrical energy is established between the two electrodes when the negative electrons at the cathode flows towards the anode and the positive hole flow towards the cathode.

The OPV solar cells are made from plastic and has the following advantages:

- Low production cost with high volume production and will be much less expensive to produce than silicon solar cells.
- Lightweight and the high flexibility of organic molecules
- High optical absorption coefficient which allows a small area of material to absorb a large volume of solar radiation.
- Very likely to be a cost-effective technology in the field of photovoltaics, based on cost projections.
- Requires less energy to produce and is therefore environmentally friendlier to produce than silicon cells.
- Very safe for use as it has no established negative environmental impact.
- Its lightweight and flexibility make it suitable for making things like bags and clothing.
- The intrinsic property that facilitates long persistent photoconductance.

The downside of the organic PV cells when compared to existing inorganic solar cells are as follows:

- Low strength
- Low stability – must be protected from ambient air to prevent degradation of the active layer and electrodes by water and oxygen.
- Low efficiency – 8.3 percent (Konarka)

Manufacture of OPV Cells

OPV cells are manufactured as an organic film in which the organic semiconductor material may be deposited in one of several ways inclusive of spin coating, vapor-phase deposition, nanoimprint lithography and vacuum thermal evaporation, similar to the methods of manufacture for other thin film solar cells.

OPV Quantum Dots

Significant work is currently being done to create organic quantum dots and one particular material and its derivatives are showing significant promise, graphene to make graphene quantum dots (GQDs).

Organic Tandem Solar Cell (OTSC)

The design intent of the organic tandem solar cells is to reduce or eliminate the energy losses associated with the single-junction solar cell unit, energy losses which are significant and which are responsible for the 32 % power conversion efficiency threshold to which these units have been confined. The major energy losses in single-junction solar cells are transmission losses and thermalization losses.

Thermalization losses. These are losses arising from high energy photons which impacting materials with a bandgap energy lower than the energy of the photon, resulting in the excess energy, (photon energy – bandgap energy), going to wastes.

Transmission Losses. These are the losses which occur due the fact that the energy levels of some photons are well below the bandgap energy of many photosensitive material, especially the wide bandgap materials, and therefore these low energy photons cannot raise the energy levels of electron hole pairs to cause the ejection and separation of electron hole pairs that would lead to the creation of electricity, these low energy photons have negligible impact, therefore their energies are considered to be lost.

The creation of the organic tandem solar cells, which are cells that are connected in series, was intended to minimize both transmission and thermalization losses and it would contain components that would utilize the low energy photons and a component that would reduce thermalization losses. The basic make up of this tandem cell is as follows:

- Utilization of a wide bandgap photoactive absorber as the first layer- this would minimize thermalization losses.

- The second layer would consist of a low bandgap organic semiconducting material – this would ensure the absorption of photons with low energy and reduce or eliminate the transmission losses

- These two layers must be connected in series with an interconnecting layer (ICL) between them. The ICL is made up of two layers, one must be a p-type hole transporting layer (HTL) and the other layer must be a n-type electron transporting layer (ETL), which when combined will act as a charge recombination layer.

While the organic tandem solar cell is designed to eliminate the transmission and thermalization losses, which they have done fairly successful, the OTSC also has other problems which must be sorted to achieve the desired outcomes. The problems associated with the tandem cells includes low intensity condition in the secondary cell, which may be caused by the ICL, this low intensity condition can impact the open circuit voltage and unfavourable energy level alignment which may also impact the open circuit voltage. The ICL is therefore very important and attention must be paid to the material selected for use in this process, with one of the most important properties required of the ICL being transparency. One other very important problem associated with the OTSC involves the processing or manufacturing issues that makes mass production of these tandem cells fairly difficult.

Quantum Dot Solar Cell (QDSC)

A quantum dot solar cell (QDSC) is a solar cell that utilizes quantum dots as the medium of collection of incident solar radiation similar to the work done by silicon in silicon solar cells and other collecting material. According to the NACK Network (2018) "Quantum Dots (QDs) are nanoparticles or structures that exhibit 3-dimensional quantum confinement which leads to many unique optical and transport properties". Another definition that grants a greater insight into the properties and behaviour of quantum dots, indicates that quantum dots are semiconductive nanoparticles, some metals also, that are usually several nano-meters in size with some very distinctive qualities inclusive of being photosensitive, photoconductive, photoluminescent and electroluminescent. These nanoparticles can be made to size to suit a particular technology requirement, such as may be required for solar cells, electronics and electronic controls systems, camera technology, quantum computing, medical imaging, smartphone battery recharging, lighting and other technologies which have the capacity to utilize nanoparticles or where nanoparticles can create new technologies to improve efficiencies or provide new ways to accomplish established functions.

The sizes of these nanoparticles can range from 2- 12 nano-meters and can be made from several different materials in group II-VI, III-V or IV-VI and compounds of these materials inclusive of lead sulphide, lead selenide, copper indium selenide sulphide, indium arsenide, cadmium selenide, cadmium telluride, indium phosphide, zinc selenide and other compounds. Carbon, Group 4A, must now also be added to this group of material as graphene, a form of carbon, is currently one of the new exciting groups of quantum dots, graphene quantum dots (GQDs), that are in development with good future prospects. All semiconductive quantum dots are relatively easy to manufacture and are excellent for use in the visible to near infrared region (VIS_ NIR) of the spectrum, 400 -2500 nm wavelengths, and because they are bandgap tuneable by size, they can be engineered to meet the needs of many different applications. Graphene quantum dots expands the range for which quantum dots can be very productive from VIS – NIR to UV – NIR, 250 -2500 nm wavelengths.

The size of quantum dots tends to be the focal point around which the technology is usually discussed however, significant importance must also be placed on the concept of quantum confinement, which is the "spatial confinement of excitons (electron-hole pairs) in one or more dimensions in a material, with each dimension given a specific assignment and function. These dimensional assignments are as follows:

- 1dimension confinement: quantum wells

- 2dimension confinement: quantum wire

- 3dimension confinement: quantum dot

Quantum confinement is usually more prominent in semiconductors, due to their energy bandgaps, and less so in metals as they do not have this energy bandgap, which causes quantum confinement in metals to be only observable at sizes below 2 nano-meters. Some other important features relating to quantum confinement are as follows:

- The Bohr's radius of the material

- The de Broglie wavelength of an electrons is a determining factor in spatial confinement as the exciton only becomes confined as its size approaches this predetermined electron wavelength in the conduction band.

- The energy bandgap is increased with decreasing particle size, the smaller the quantum dot the higher the bandgap energy

- Quantum dots are bandgap tuneable by size.

- The absorption and the emission of photon energy occurs at specific wavelengths which are directly related to the size of the quantum dots.

TYPES OF QUANTUM DOTS

In general, there are three types of quantum dots, core-type quantum dots, core shell quantum dots and alloyed quantum dots.

Core-type Quantum Dots. These quantum dots are made up of one material only with a uniform internal composition and the materials used to make these quantum doats may include selenide, telluride or sulphide compounds of lead, cadmium, silver or zinc. The optical and electronic properties of these semiconductors may be varied by increasing or decreasing the size of these nanoparticles.

Core and Shell Quantum Dots. The properties of semiconductor quantum dots, inclusive of efficiency and brightness, may be varied by enclosing the core dot into a higher bandgap semiconducting material and these dots are made up of one core material with a shell around it are usually referred to as core and shell quantum dots. These core and shell quantum dots may be of cadmium selenide on the inside and zinc sulphide on the outside and many other such combinations.

Alloyed Quantum Dots. The alloying of two semiconductor material with different bandgap energies will provide another opportunity to tune the properties of quantum dots, as the alloys formed will display properties that are very different from their component parts and from their bulk parent material.

Advantages of Quantum Dots

According to NREL the most important factor that must be dealt with today in the solar energy industry is the ability to convert more solar energy into usable electrical energy when utilizing non-concentration solar energy technologies and systems. Currently there is a limit on the amount of energy that can be recovered from a non-concentrated solar energy system and this limit is based on the efficiency at which current solar system can convert the incident solar energy into electricity, the conversion efficiency, and based on work done by Shockley and Queissar in 1961 (Nozik 2001) the highest conversion efficiency that could be achieved was 32 %. This limitation is a function of the mechanisms that existing solar cells use to absorb photon energy, liberate the electrons that create the flow of electricity and the portion of the photon energy that is lost . In standard silicon solar cells one photon of energy liberates only one electron-hole pair at-a-time, with the release of a higher number of electron-hole pairs requiring the equivalent number of photons and similarly, in the new generation organic solar cells, one photon of light will only release one exciton. According to Prezhdo (2008) from the research work conducted up to that point with quantum dots, it has been established that quantum dots are more productive in this way as one photon absorbed by a quantum dot could release three electron-hole pairs, multiple excitons (MEs). This has been confirmed by NREL (2013) "Today's solar cells produce only one exciton per incoming photon, but the "multiple exciton generation" (MEG) effect of quantum dots promise to wring more energy out of each photon", clearly indicating that QDs have the capacity to increase the conversion efficiency to well over the existing 32 % limit and into the 66% range (NREL) making the quantum dot solar cells a significant game changer.

According to NREL there are three types of quantum dot solar cells configurations that are currently being worked on by several research institutions and these three types are photoelectrodes composed of quantum dot array, quantum dot-sensitized nanocrystalline TiO_2 solar cell and quantum dot dispersed in organic semiconductor polymer matrices.

Photoelectrode Composed of Quantum Dot Arrays. The quantum dots in these photoelectrodes are arranged in a structured 3-D array with relatively small spacing between the quantum dots to facilitate the development of strong electronic bonds and minibands that will allow long-range electron transport. In this configuration the detached, well defined 3-D miniband states would produce the effect of slowing down the carrier cooling and this slowing down of the cooling would facilitate the transportation and collection of hot carriers that would allow the generation of a much higher photo-potential in a photovoltaic cell, in which quantum dot 3-D array is being used as the photoelectrode.

Quantum Dot-Sensitized Nano-Crystalline Titanium Dioxide Solar Cells. Titanium Dioxide is a semiconductor material that is a natural occurring crystal, that has become one of the most important photocatalyst. As a photocatalyst titanium dioxide has found many uses in industry inclusive of, as a source for metallic titanium, a photocatalyst for solar cells, an important alloying element that imparts high tensile strength, high corrosion resistance and high temperature resistance to metals, a paint pigment, a bleaching agent and for use in cosmetics. Titanium dioxide is also being highly utilized in the field of energy, specifically solar energy as it can be used in combination with other elements to improve the collection of photon energy and the release of excitons which consequently creates electrical energy, two such combinations are with dye and quantum dots. Titanium dioxide is a wide bandgap semiconductor that absorbs ultraviolet (UV) light, however, its ability to convert the broader spectrum of incident solar radiation into other forms of energy is fairly limited due to its high bandgap energy. Experience, however, has shown that the spectrum absorbed by the titanium dioxide could be broadened by the use of sensitizers inclusive of dye and quantum dots.

To achieve the require sensitization the quantum dots must be attached to the titanium dioxide and this can be achieved using any one of two methods, chemical bath deposition (CBD) and linker chemistry both of which have been found to be fairly efficient. In the CBD method the quantum dots are deposited or grown directly onto the surface of the titanium dioxide, while in the linker method the quantum dots are attached to the surface using other compounds which will impacts the capacity of the cell to collect incident photon and create electricity. Extensive work with both dye and quantum dots has shown that the use of quantum dots in this way is not an improvement on the use of the dye, as they both achieve very similar results.

Quantum Dots Dispersed in Organic Semiconductor Polymer Matrices.

In this configuration the Quantum Dots are dispersed in a blend of organic semiconductor polymer matrices that have been established through significant research to facilitate the collection of incident solar radiation, the creation of excitons and the transportation and collection of charges. This organic semiconducting polymer also has the capacity to facilitate the ejection of multiple excitons for the creation of electricity and the combination of the QDs with the semiconducting polymer will enhance the capacity of organic semiconductive polymer. This combination has been used in many applications inclusive of providing enhanced lighting for painting and marking and for biomedical purposes.

CHAPTER 10

An Innovation in Solar Energy Technology

Solar energy is currently utilized by collecting it in what has become the accepted standard forms for the different solar energy systems inclusive of photovoltaic (PV) systems, concentrated solar power (CSP)energy systems and solar heating and cooling systems (SHC). For the PV systems this involves the creation of solar cells from photoconductive materials, the integration of many of these cells to form a module and the integration of many modules to form arrays that may be utilized to generate significantly large volumes of electrical energy. In the CSP systems several types of mirrors or reflectors and accessories are utilized to concentrate and reflect incident solar radiation onto heat conducting receivers or collectors. The heat collected in this manner is utilized to raise the temperature and pressure of heat transfer fluids (HTF) that may then be utilized to generated high temperature and high-pressure steam that will be suitable for driving a turbo-generating unit to produce electrical energy. In the SHC systems the solar energy is collected by a group of flat, inclined heat exchangers that utilize a heat transfer fluid that is circulated through the system which is used to heat water for domestic or other uses or the heat exchanger may heat water directly for direct consumer consumption. These established methods of collecting solar energy while relatively successful, have significant problems and inefficiencies that impact the environment in many negative ways and other methods of collection may need to be found to eliminate the related problems.

All solar collectors are currently manmade systems that require enormous effort and cost to design, fabricate, install and to dismantle at the end of the useful life of the equipment, as typically there are no natural systems that can create and transfer electrical energy for direct use by a human population. Currently, while there are natural systems that can create electrical energy, such as lightening, voltage from electric eels and other animals that discharge electrical energy, no way has yet been found to harness these natural sources of electrical energy. The ability to harness such naturally produced electrical energy would provide the optimal energy solution, if these natural systems could be multiplied or replicated to produce significant quantities of electrical energy to meet a portion of the world energy demands, since these natural energy producing systems are a natural part of nature and are very environmentally friendly as they have no known negative impacts on the environment or specific ecological systems. Since the harnessing of naturally produced electrical energy is not yet feasible it becomes critical that efforts are made to ensure the best possible environmental outcome of producing electrical energy by creating electrical energy generating systems that closely resembles natural system. To achieve such outcomes will require the study of natural systems, their natural architecture and structure in order to facilitates the creation of imitation natural system that can produce electrical energy in an environmentally safe and friendly way.

Currently there are at least three systems that harness natural energy to produce electrical energy and these three systems utilize the energy of the natural wind, the natural radiation of the sun and the natural power of water, both fresh and saline, all renewable energy systems. The harnessing of all these natural energies requires the utilization of well-designed mechanical systems all of which have significant impact on the natural environment based upon the volume of land required to construct, install and operate these systems, the locations chosen for these installations, the material utilized to construct, their impact on natural hydrology, their impact on natural flora and fauna, their impact on life sustaining agriculture and aquaculture and their consequential impact on human life. Some of the problems associated with these renewable energy systems are as follows:

- **Wind**. Electrical energy may be produced by wind harnessing systems both on land and offshore in the oceans and the harnessing of wind requires the utilization of large volumes of land, both onshore and offshore. The creation, installation and operations of wind harnessing systems for extended period of time usually requires the clearing of large volumes land, inclusive of trees and plants where virgin land is used. The large volumes of land are required for the location and installation of the very large wind turbine machines, the creation of access roads and service areas, and these clearing activities usually destroy significant natural habitat for many species. The operation of the wind plants will also significantly impact the lives of reptile, amphibians, birds, insects and other flying animals such as bats whose natural habitat and feeding locations would have been impacted by the location of the plant.

- **Water**. The utilization of the potential energy of water in any one of several hydropower systems will require much effort that could negatively impact the environment in which the necessary machinery is located, in several different ways. In general, land-based hydropower systems required the displacement of human populations, the destruction of large volumes of trees, the submersion of large acres of land to create the necessary dams, lake or reservoirs to create required potential energy in the water and in the process destroy many historical site, many species of plants and animals and a major disruption of the natural hydrological systems. The systems installed in middle of rivers (in-stream installations) or in the oceans also negatively impact the place in which the necessary machinery is located as they will impact the natural flow of water, the natural habitat of aquatic flora and fauna and eventually the humans that are dependent upon the particular bodies of water. Historical data has also shown that many ecological systems that are dependent upon the flow of the body of water will also be negatively impacted in many unexpected ways such as the destruction or disappearance of a species of plants that are the natural food for a species of deer which are the natural prey for a species of carnivores such as wolves. The instream machinery, turbines, will also negative impact the natural flow of the stream and impact the hydrology and instream aquatic flora and fauna and these same negative impacts also affects the natural wave patterns and the aquatic flora and fauna of the ocean when heavy machinery is placed in the ocean for wave and tidal energy systems.

- **Solar Energy Systems**. There are currently three solar energy systems inclusive of Photovoltaic (PV), Concentrated Solar Power (CSP) and solar heat and cooling (SHC) systems and two of these systems, PV and CSP systems, when utilized for utility scale power plants, requires the exclusive use of large volumes of land. This exclusive use of large volumes of land will usually require the clearing of green fields, the levelling of existing agricultural lands and the exclusion of all other possible usage of the land, especially agricultural use, when utilizing existing PV solar system technologies and probably even new technologies, if the system of collection is to be maintained in its current format of cells, modules and arrays built into flat, inclined systems. The same also applies to CSP technology which is not expected to change much in the near future, especially in its demands for the exclusive use of high volumes of land. The concentration of solar energy beams created by CSP systems also has the capacity to destroy birds, flying insects and bats all of which may be very important in agriculture for the pollination of vital crops. This impact plus the demand for the exclusive use of large volumes of land doubles the negative impact, as solar energy plants are now directly competing with agriculture for the use of the same land, good, level, easy to use land, all around the world.

The many negative impacts caused by the mentioned renewable energy systems require that all effort be made to reduce or eliminate these impacts if these renewable energy systems are to continue producing and increasing their positive impacts upon the environment, especially on the maintenance of conditions in the atmosphere to prevent the Earth from reaching the tipping point in climate change, the 1.5^0 C temperature rise. The effort to reduce their negative impacts should be focussed on utilizing systems that incorporate as many natural or natural like elements into existing systems designs, to reduce their negative impacts upon the environment. The many improvements being made in the field of solar energy technology makes this renewable energy sector a significant target area for improvements and to date significant amount of work has already been done to mimic natural technology, especially with the design of the dye sensitized solar cell (DSSC), a solar cell which was said to be a direct attempt to replicate the natural process of photosynthesis, for use in the creation of electrical energy from solar photovoltaic systems.

DSSC and Photosynthesis

Nature and natural biological systems have in the past inspired many great leaps forward in science and technological advances and some of nature inspired advances include the invention of flight by humans and hovering by helicopters, both inspired by the flight and hoovering of birds and insects, nature's best fliers. This copying of nature however, did not start with birds as many plants have inspired medicines for centuries, the arts have also been inspired by both plants and animals and it is very likely that nature will continue to inspire growth and changes in all areas of science, technology, art and medicine for a very long time to come. The natural process of photosynthesis inspiring the design of a light gathering and conversion device, in this case DSSC, falls naturally into the trend just described and could prove to be the biggest boon to renewable solar energy systems and to the world's supply of clean energy at a not too distant point in the future. Interest in copying and utilizing the photosynthesis electro-chemical process began in the 1970s and this interest later led to the development of what is now called dye sensitized solar cell (DSSC).

A brief description of the process of photosynthesis is as follows:

- A photon of light is absorbed in the thylakoid membranes of chloroplasts
- The energy of this photon excites a specific protein chain and raises the energy level of one electron in chlorophyll pigment.
- The photon energy also splits water into a positive hydrogen ion(H^+) and liberates oxygen (O_2)
- An electrical current is produced by the electron transfer that take place in the chain of proteins and this electrical current is used to drive other systems within the leaves of the plant.

The processes in the DSSC are as follows:

- Light is absorbed by the dye that is covering the surface of titanium dioxide (TiO_2) crystals
- The electron ejected from the dye by the photon energy is injected into the conductance band of the titanium dioxide (TiO_2)
- The flow of electron from the dye leaves a positively charged hole in the dye.

- The electron displaced from the dye is replaced by an electron from an electrolyte that forms a part of the DSSC.
- The electron completes the circuit and returns to recharge the electrolyte and continue the cycle.

According to Adedokun et al (2016) The operation of a DSSC cell is very similar to photosynthesis in leaves and the common elements are as follows:

- Photosensitizer. Dye in DSCC versus chlorophyll in Photosynthesis.
- Electron Acceptor. Titanium Dioxide in DSSC versus Carbon Dioxide (final acceptor) in photosynthesis
- Electron donor. Dye in DSSC versus Chlorophyll specie in Photosynthesis
- Electrolyte. Electrolyte in the DSSC versus Water in photosynthesis.
- Transparent protective layer. Transparent glass substrate versus natural transparent plant substrate, cuticle.

The development and creation of the dye sensitized solar cell, an imitation of plant life, was quite innovative and this innovation came with a great hope of improving solar energy collection and conversion into usable electrical energy and thereby helping to reduce the dependence upon the use of carbon rich fossil fuels and reducing the impact that carbon has our climate and life on earth. But this innovation, DSSC, only imitated one portion of the greater possibilities that may be achieved by imitating a larger portion of plant life, as plants have so much more to offer based upon their long existence in making life on earth liveable by their utilization of solar energy. The greater possibilities obtainable by imitating more of plant life includes the possibility of producing more clean, renewable electrical energy, the stabilization of soils, the reduction in soil erosion by wind and water and the provision of habitat for many species of flora and fauna.

Plants were among the earliest collectors and users of solar energy on planet Earth, as they have been collecting and utilizing solar energy for more than 400 million years, in the Devonian period, long before human existence and the human need for sources of energy. This time period, 400 million years, saw beginning of the evolutionary life and development of plants, this time period is as long as plants have existed on planet Earth, and these plants have continued to evolve and adapt to ensure that they could continue collecting solar energy even in the most inhospitable of locations and situations. Plants can be found on land in good arable soil, in semi-arable soils, in rocks, in deserts, in water, under both fresh and saline water, in freshwater and saline swamps, in deep crevices and ravines, just as long as sunlight and water are available, sometimes just a little water, sometimes just little water. A better understanding of plants, especially the giants of the plant kingdom, trees, can be obtained with a general review of the natural architecture of trees and this review must be carried out with a view to identifying how more of the features of trees can be utilized in the effort to provide more clean energy for human consumption.

Natural Architecture of Trees.

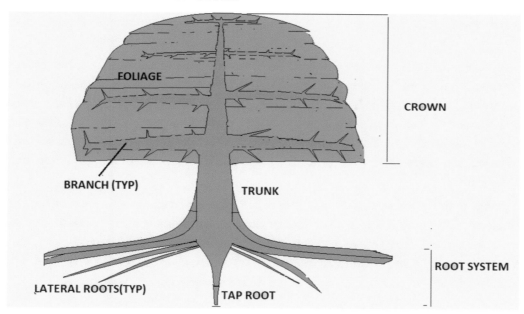

Fig 10.1 Typical Tree Architecture

The natural architecture (NATARCH) of trees is as shown in Fig 10.1 above and this diagram is only a general representation of trees as it does not represent all tree, as trees have evolved with many different shapes, leaf type, root systems, trunk sizes, heights, flowers, fruit types, and even the average length of their lives. The important points to note from the diagram is that all trees have roots, a trunk, a crown, branches, foliage and a bark and the importance of each of these is as follows:

- **Roots**. The roots are one of the most important parts of a tree as it performs several critical functions inclusive of connecting the tree to sources of water and nutrient, anchoring the tree firmly in and on the earth, stabilizing the tree during wind events, acting as a storage area for the tree and most importantly acting as the foundation, the most important structural support for the tree.

- **Trunk.** The trunk serves several very important functions inclusive of, structural support for the crown, the only connection between the crown and the ground, providing the required elevation for the crown to ensure that the leaves get enough exposure to sunlight, acts as the main conduit that delivers water and nutrients to the rest of the tree from the roots up and delivers plant foods from the leaves down to the roots of the tree. Most trees have only one trunk, but quite often during its early development some trees, depending on its specie, may divide near the roots and end up with what appears to be more than one trunk.

- **Crown.** The crown consists of the main branches and sub branches or limbs, twigs, leaves or foliage and fruits and it acts as the manufacturing center of the tree as it is the location where the important elements of water, nutrient, carbon dioxide and sunlight come together in leaves to complete the process of photosynthesis to produce the necessary plant food that helps the tree to grow, produce fruits and keep itself alive. The most important function of the crown, however, is the elevation and structural support that it provides to the leaves to allow them to be exposed to the optimum level of sunlight that they need to produce optimally.

- **Branches.** Each tree has many branches of different sizes ranging in size from the lowest to the highest main branches which are directly attached to the trunk of the tree, sub branches or limbs which grow from the main branches, down to the twigs which are the smallest branch partitions on the tree. The main function of these branches is to provide structural support for those portions of the tree that will be bearing fruits and to expose the leaves of the trees to the widest possible variations in angles to ensure that all leaves receive adequate sunlight for production. The lower main branches are usually more robust than the branches at higher levels with the sizes decreasing with each step up and away from the trunk with twigs being the smallest.

- **Foliage.** The foliage are the leaves of the tree and they are at the centre of life for all trees irrespective of the tree species, size, shape, locations of origin, or the type of fruit that the tree may produce. Most trees are covered all around with leaves, 360 degrees, from the trunk up to the highest point of the crown and each species of tree will also have different type of leaves, leaves of different shapes, sizes, different shades of green and the leaves may also be connected in many different ways to the trees. Some of the many different types of leaf connections are as follows, alternate, odd pinnate, perfoliate, tripinnate, bipinnate, opposite, rosette, unifoliate, even pinnate, peltate, trifoliate/ternate and whorled. The leaves also come in different sizes, surface finish, edge types and length of life which varies from one year to multiple years, all dependent upon the species of the tree. The main functions of the leaves are to act as solar collectors- absorbers of photon energy and the facilitators of the ejection and transfer of electrons, the storage area - for the collection of carbon dioxide, water and nutrient, the processing center - where water, nutrients, carbon dioxide and sunlight are brought together in correct proportions to complete the process of photosynthesis to produce plant food and fruits and as the facilitators of transpiration an important process for all trees.

- **Bark.** The bark is the skin of the tree and it covers every section of the tree starting from the roots up to the twigs and its main function is the containment and protection of the vascular system of the tree.

A comparison of the physical structure of the various components of the tree and a standard solar system will show a similarity with the components of existing photovoltaic solar energy systems as follows:

- **Collector.** Crown of the tree = PV Solar Modules/PV Solar Arrays
- **Structural Support.** Trunk of the tree = frames for the physical supports of solar panels
- **Conduits.** Bark of the tree protecting vascular system = conduits to protect cables and wires to supply controls and to deliver the developed electrical energy.
- **Solar cells.** The leaves on each tree = to the solar cells which make up PV modules, the arrays and the overall collector.
- **Glazing.** Protective transparent layer of each leaf(cuticle)= protective transparent glazing on each panel or solar cell.

From the comparisons above and other observations, trees could be said to have several advantages over man made solar collectors and these advantages include the following:

- Trees are biodegradable and while it returns the carbon to the environment when it is decaying, decomposing trees serves many important functions such as providing habitat and food for many forms of small plants and animals which are very important for the environment and many ecological systems and therefore decaying tree will do much less harm to the environment than any manmade systems at the end of their useful existence as solar systems.
- Trees in general have much longer lives than solar panels which have a useful life of only 20-30 years and this longer life will ensure that trees will return carbon to the environment less frequently than the manmade system.
- Trees collect and utilize both direct and diffuse radiation and will automatically collect more incident solar radiation than the standard solar collector (PV, CST or SHC) which mostly collect direct radiation.

- Trees, while fixed will never require rotation as they have leaves all around their perimeter to collect energy on all sides, beginning from sunrise to sunset.

- Trees collect solar radiation for longer periods of time, 8- 12 hours, than most standard manmade solar collectors which collect solar radiation for only an average of 5-6 hours of daylight each day.

- Shading does not impact some plants in the same manner that it does solar panels as trees can grow their way into the sunlight and some trees and plants in fact thrive under the shadows created by the canopy of larger trees.

- Trees do not require the exclusive utilization of land as the standard solar systems do.

- Trees form a critical support system for many species of flora and fauna, which manmade solar systems do not.

- Each tree will have thousands of solar collectors (leaves) available during all daylight hours facing many different directions, collectors that are always ready and available to collect solar energy from both direct and diffused solar radiation, from sunrise to sunset.

- Trees do not require flat arable lands for growth and production as most PV solar energy farms do.

Trees have utilized their natural architecture (NATARCH) to support the natural collection and utilization of solar radiation to grow fruits and propagate themselves and this natural architecture of the trees (NATARCH) could probably be adopted to similarly collect and convert solar radiation to produce some good results in the field of PV solar energy. While the natural architecture (NATARCH) of the tree is great for trees, there may also be some significant disadvantage to the utilization of the natural architecture of trees for solar energy systems, the most significant of which are as follows:

- Many species of flora and fauna utilize natural trees as their natural habitat and may attempt to utilize the manmade structures in a similar manner and consequently negatively impact the operation of any PV solar systems based on this design.

- The overall cost to design and construct a NATARCH solar tree with all of its essential parts and accessories may be too high.

- The susceptibility of the NATARCH solar tree to natural weather events, such as snow and wind events.

- Research done at Michigan State University and elsewhere, indicates that the basic design of leaves and the systems that supports plants, make leaves less efficient at capturing solar energy than solar cells and also that the process of photosynthesis is less efficient than solar cells at collecting and converting solar radiation into a useful force.

The many disadvantages of the natural tree architecture highlighted above should not, however, prevent the complete review of the natural architecture (NATARCH) of trees with a view to utilizing it in the production of solar energy, but should instead provide an impetus to complete the necessary work to investigate the best facets of plants, especially the giants of the plant kingdom, trees, that may be useful in supporting PV solar systems to create electrical energy, based on their long history of collecting and converting solar radiation. There are many species of trees with many different characteristics inclusive of different growth characteristics, types of soils and locations that are most suitable for them, different heights, different branching characteristics, different leaf sizes and types, different timber characteristics, different levels of tolerance to water, salt and dry arid conditions. Even with their many differences, most trees have one thing in common, with the exception of a few plants, they all have leaves that are responsible for the collection of solar energy that gives them life and this characteristic is the most important one, relative to the rest of their architecture, in the collection of solar energy.

The relative importance of leaves therefore makes it imperative in the quest to copy the natural architecture (NATARCH) of trees for PV solar energy purposes that the focus must be on the crown and not on the trunk and root support system, as in manmade PV solar energy system the roles of the structural support become secondary. In the natural tree the crown consists of branches, limbs, twigs, leaves and fruits, all of which will significantly impact the operations of the natural tree and will likewise the operation of a NATARCH solar tree, because similar to the support given to natural trees, the branches, limbs and twigs of the NATARCH solar tree will also act as structural supports that positions the solar collectors to ensure optimum incident solar radiation collection, as supports for the control systems that will deliver control signals to the collectors and as supports for the conduits that carry electrical wires that transfer generated electrical energy to take off points. The leaves on the NATARCH solar tree will also be acting in a similar manner to leaves on natural tree, as the solar collectors, that will absorb the incident photons of light, support the ejection of electrons that will create the flow of electricity and support the transfer of electricity to the terminals. The leaves of a natural tree have a basic infrastructure that is very important in the operation of each leaf and the basic infrastructure is as shown in the diagram in the figure below.

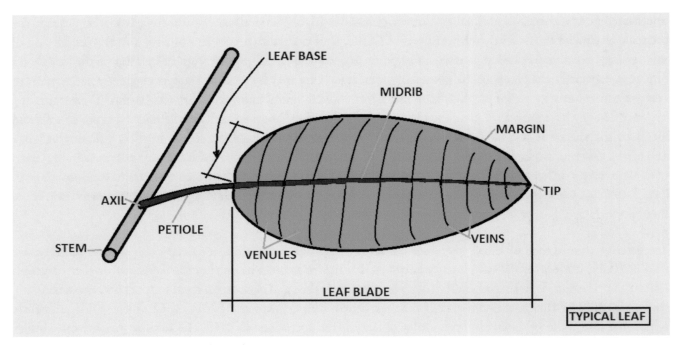

Fig. 10.2 Diagram Of A Typical Leaf

This diagram does not represent the shape of all leaves but is just one of many shape and leaf shapes vary from the shape in the Fig 10.2 above to many other shapes inclusive of circular, oval, triangular and oblong. Leaves also vary in sizes, length and width, and can have dimensions that vary anywhere from over 2 m for a single leaf of the giant taro, (The Living Rainforest) or 3.3 m for the *Gunnera manicata* to the miniscule sizes of the Wolffia aquatic plants and the 0.6 centimetres length leaves of the giant pine trees. These leaves irrespective of size, all perform the same basic functions that are essential to the life of the plant of which they are a component. Another dimension of great interest is the thickness of leaves which also varies with the species of each plant and the environmental conditions in which the plants are grown. The thicknesses of the many leaf types and species can vary anywhere from 11 centi-meters of the *Gunnera masafuerae* to the thinness of the leaves of the Wolffia aquatic plants which are less than one milli-meter in length and a much smaller thickness. The thickness of leaves has significant impact on the behaviour of the leaf and on the process of photosynthesis, as indicated by Pauli et al (2017) "Leaf thickness largely determines the length of the optical path of light through a leaf and the number of anatomical features (eg cell walls and chloroplasts) that either reflect, absorb or transmit light". The average thickness of leaves varies by specie with rice having an average thickness of 331-307 micro-meter (Flag leaves) and tobacco 235 – 250 micro-meters, and the relative thickness or thinness of the leaves are however, only important in demonstrating the scale at which many very important activities are taking place in the leaf, the micro to nanoscale.

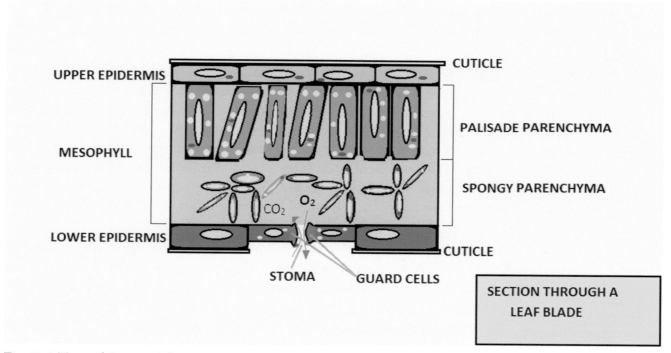

Fig 10.3 Typical Section Through A Leaf Blade.

The basics components of a leaf and their functions are as follows:

- Cuticle. The cuticle, formed by the epidermis from non-cellular material, is the outermost surface of the leaf and this non-cellular material can normally be found on several areas of the plant inclusive of the surface of leaves, fruits and plant body. The cuticle acts as protection against extreme temperatures, excessive water loss, UV radiation, droughts, chemical attack, mechanical injuries, pathogens and or pest infections. The basic components of the cuticle are usually wax, cutin, terpenoids and flavonoids, with the terpenoids and flavonoids both having antifungal properties.

- Epidermis. Each leaf has an upper and a lower epidermis, one on the upper side of the leaf and the other on the lower side and they are the outer layers of the leaf. The main function of the epidermis is to create the cuticle and combined they act together to prevent excessive water loss and to prevent the intake of harmful substances such as pathogens. They also regulate the intake and outflow of gases, carbon dioxide and oxygen, through the stomata on the lower epidermis, the lower side of the leaf, as shown in FIG 10.3 above. The epidermis is usually only one cell layer thick except in either extreme conditions of either cold or heat, where the epidermis is usually several cell layer thick to prevent excessive water loss from transpiration.

- Mesophyll. The mesophyll is sandwiched between the upper and lower epidermis and it contains two very important components, the palisade parenchyma and the spongy parenchyma both of which contain chloroplasts and are very important to the process of photosynthesis. The palisade parenchyma is usually right beneath the upper epidermis, are shaped like columns which are tightly packed together, contains chloroplasts and are efficient light absorbers. The spongy parenchyma cells are usually loosely packed, contains significant amount of air pockets to facilitate the gas exchange processes and are usually covered with a thin layer of water in which the gases are dissolved during the gas exchange process. During the photosynthesis process the carbon dioxide is absorbed by the wet cells and the cells later releases the oxygen.

- Stomata. These are microscopic pores in the epidermis that are responsible for the intake of carbon dioxide to and the release of oxygen from the spongy parenchyma. These pores are normally located on leaves but are sometimes located on stems with the majority of the pores located on the lower side of the leaf for land plants and in aquatic plants the majority of the stomata are usually found on the upper side of the leaf. The stomata are usually created and controlled by the guard cells, to regulate water loss due to transpiration, carbon dioxide inflow and oxygen outflow. The stomate will open in response to light and will close whenever there is drought stress, excess carbon dioxide, low humidity and ozone.

- Guard Cells. Guard cells are a pair kidney or bean shaped cells between which stomata are formed and these cells are mostly located on the lower epidermis for land plants and on the upper epidermis for aquatic plants. These guard cells are specialized cells whose external structure is made up of polysaccharide-based wall polymers that has high strength and are fairly elastic. The guard cells are generally controlled by the brightness of the available light, with bright light causing the guard cells to take in water by osmosis and as this water level increases the cells become swollen and open the stomata, the stomata close when the decreasing light levels cause the guard cells to lose water and become soft, closing the stomata.

Collection Efficiency of Different Leaf Types

The leaves of all plants, land and aquatic, are responsible for the collection of the solar energy needed by each plant type with each plant species having very different type of leaves to meet their specific needs, needs which differ from specie to specie and the size and type of leaf does not reflect the size and type of plants. We could therefore have small plants such as the water lily and giant taro having oversized leaves while many giants of the forest have very small needle like leaves which raises the question of which leaf types are more efficient at collecting and utilizing solar energy for plant development, growth, fruit production and maturity. An examination of the importance of leaf types, shape, size and thickness needs to be done to determine the size and type of leaves that may be suitable for consideration in the development of any NATARCH solar energy collector. As may be attested to by many botanists, a leaf's size and shape is very important in relation to their ability to perform their basic functions inclusive of photosynthesis, transpiration and thermoregulation, and the size and shape of the leaf may be affected by environmental condition inclusive of altitude, temperature, annual rainfall, type and fertility of the soil, typically the size of leaves will tend to decrease with an increase in altitude, lower mean annual temperatures and low levels of rainfall. According to Zhang et al (2021) a leaf's size and shape will affect its ability to collect solar radiation, perform photosynthesis and affect plant growth and they also indicated that larger leaves collect more solar radiation, direct and indirect, and does a better job with the process of photosynthesis.

When comparing leaves and solar cells it must be obvious that leaves quite clearly perform a lot more functions than solar cells do, as the solar cells performs only four functions, the collection of incident solar radiation, the liberation of electrons from a donor source, the acceptance of the electrons by the accepting agent and the flow of electricity through terminal electrodes. The performance of fewer functions than a leaf, places solar cells at a distinct advantage when efficiency comparisons are made with leaves and their capabilities to convert photons of light to electrical energy. A better comparison of solar technologies, natural and manmade, could be made if the additional functions carried out by leaves are eliminated and natural leaf solar technology is allowed to compete on an even footing with other solar technologies while still being supported in their natural settings or architecture. This natural setting would include a "tree" complete with "roots", "trunks," "branches", "limbs", "twigs" and "leaves", thousands of leaves, and any other naturally required attachments to recreate the natural architecture.

Design Initiation for the NATARCH Solar Tree

Unlike natural trees which grow in the soil from seeds or other propagation methods, the NATARCH solar tree must be designed and built to meet standards, as may have been established by government and other bodies, and the most important elements to be considered in the design process for the NATARCH solar tree are the following, solar collection system, structural support for the collection system, structural support for the overall structure of the NATARCH solar tree, electrical support system, control support systems, connectivity systems, protection systems and interconnection systems.

Basis of Design for the NATARCH Solar Tree

The design of the NATARCH Solar Tree shall be arrived at from the following:

- The utilization of technology that will best facilitate the concept of imitating or mimicking the natural architecture of trees.

- The utilization of energy efficient third generation solar technology for the design and creation of the solar cells.

- The utilization of organic components in every portion of the solar tree, where possible, starting with the solar cells.

- The utilization of environmentally sound design concepts that will facilitate the integration of these solar trees into the natural environment and the design shall also facilitate multiple usage of the land to include activities such as solar energy installations, agriculture, education, recreation or other desirable and feasible activities.

- Ensuring that the design of the NATARCH solar tree will provide significant shade which has the capacity to prevent drying out of the land and thereby facilitate development and growth of agriculture in the shadows or near shadows.

- The utilization of the best available electronics in the construction and operation of all of the support systems, inclusive of inverters, converters, charge controllers, MPPT, protection systems and the necessary sensors.

- The utilization of the best available battery energy storage system (BESS) technology for the solar energy system.

- The utilization of organic material on the external surface of the trunk, below the solar power section, to provide insulation, the capacity to absorbs some water, and the capacity to grow some fungus (similar to tree bark) to facilitate the creation of habitat for the many species of insects that could live on and around these trees without disrupting the solar power generation activities and operations.

- The utilization of material with the required structural strength and integrity to ensure that all components will be strong enough to withstand the natural elements inclusive of all extremes of wind, rain, snow and hail.

- The utilization of new or existing methods, material and technology that will ensure that the cost to produce and operate these NATARCH solar tree systems is kept within affordable and practical ranges. The desire to contain cost at every stage of development is of critical concern as cost may determine if the project is to survive just being a nice concept to become a reality, even though the economic production of electrical energy from NATARCH solar trees is not yet the goal at this stage of this project.

- The utilization of conduits on the inside to ensure the safe transfer of the electrical energy created to the ultimate take-off point and the supply of necessary control signals to the power plant from any external operation center or the electrical grid.

- The utilization of the necessary systems and equipment to provide the solar tree with the ability to clean its leaves or collectors when necessary and these necessary systems may include the utilization of water sprinklers nozzles, pumps, piping, valves, electrical and controls systems

- The utilization of the necessary access systems and equipment, to provide necessary access to reach all solar collection surfaces and accessories from the ground to the top of the NATARCH solar tree, to facilitate necessary repairs and maintenance.

All of the above provide the general basis to design the overall NATARCH solar tree systems only, therefore a more specific basis of design is also required for each element of the tree, inclusive of the general structural framework, the collectors (leaves), the structural supports, the control systems, electrical systems, the battery storage systems, the cleaning systems and the access systems.

The Framework of The NATARCH Solar Tree

The framework of this solar tree is very important to achieve the set goals of optimizing the collection of incident solar radiation and the concept of utilizing the land for more than one purpose, the design of the solar tree framework shall be according the following:

- The utilization of adequate dimensions on all sections of the framework to ensure that the operational systems of the solar tree is high enough to facilitate the practice of agriculture and any other desirable activity beneath the canopy.

- The utilization of the best geometric shape for the crown of the solar tree that will maximize the exposure of the solar collectors each day.

- The utilization of frames at every level of the solar tree that will facilitate the exposure of the maximum number of collectors at any one time, with every hour of daylight.

- The optimization of the dimensions of the solar tree to facilitate the maximum number of collectors that will allow the optimization of the amount of energy that may be collected by the tree.

- The utilization of mechanical joints and connectors that will facilitate the timely construction of all the frames and support systems of the NATARCH solar tree.

- The utilization of mechanical and other connection systems that will facilitate the attachment of the leaves.

- The utilization of structural systems that will optimally stabilize the crown of the NATARCH solar tree.

- The utilization of the best available electrical infrastructural support systems to facilitate the necessary conduiting and wiring to make the electrical and control systems function as they were designed to.

- The utilization of the required structural support systems to stabilize the NATARCH solar tree in the ground, the root structure.

Collectors for the NATARCH Solar Tree

The collectors for the solar tree shall be designed using the following concepts and guidelines:

- The collector should be designed in the shape of a natural leaf or a close approximation of this shape.

- The size of the leaf to be utilized shall be relatively large, the size that is considered most efficient in nature.

- The photoconductive element to be utilized in the collector shall one of several third- generation solar technologies which incorporates organic polymers.

- The utilization of transparent collector material to ensure the best operational capabilities for the NATARCH solar trees.

- The utilization of the required electronics in each leaf to reduce or negate the natural impact of shading, power optimizers.

- The utilization of the optimum quantity of leaves to ensure collection of the maximum amount of incident solar radiation each day and the number of leaves shall be similar to the quantity on a natural tree which has approximately 200,000 leaves or the number of leaves to be utilized on each NATARCH solar tree shall be equivalent to the surface area required to produce the amount of power that is to be generated.

- The utilization of material that demonstrates the phenomena of large and persistent photoconductivity (LPPC).

- The utilization of a collector material and technology that is sensitive to and can absorb and convert both direct and diffuse solar radiation to produce the desired photovoltaic effect.

- The utilization of flexible high strength material to ensure the durability and longevity of each collector and its connections, to ensure a minimum lifespan of 30 years.

Structural Supports for the NATARCH Solar Tree

The design of the structural supports inclusive of trunks, branches, limbs and twigs shall be based on the following:

- The utilization of an existing or new lightweight, flexible material with high strength inclusive of in torsion, tension and compression.

- The utilization of material with the capacity to resist corrosion, wear and tear.

- The utilization of material that will facilitate the use of existing mechanical joining methods or other available joining methods.

- The utilization of material that will facilitate the proper grounding of the whole solar tree system if necessary.

- The utilization of material that will allow the bonding of insulating or protecting material where required.

Control System for the NATARCH Solar Tree

The controls systems for these trees shall be a function of the following:

- The utilization of the latest electronic technology to optimize the operation of this solar tree system.

- The utilization of the optimum level of electronic controls to suit the complexity of the systems in operations, single units or multiple units.

- The utilization of appropriate electronics control systems to ensure safe and efficient connection to and from residential, commercial and industrial facilities and the electrical grid.

- The utilization of appropriate electronic control technologies inclusive of the MPPT, Charge Controller, DC-DC converters, inverters, communications technology and power optimizers or equivalent technologies to ensure optimum operation of the solar energy power plant and transfer of the produced electricity to all facilities and the electrical grid.

- The utilization of the appropriately sized electrical conductors, equipment and accessories to support flow of control signals.

Electrical System for the NATARCH Solar Tree

The electrical system responsible for the collection and transportation of the electrical energy generated is of optimal importance to the solar tree and should be designed as follows:

- The utilization of the appropriately sized insulated electrical conductors and accessories at the terminals of each collector

- The utilization of the appropriately sized grounding conductors, rods and accessories for each collector.

- The utilization of the appropriately sized insulated electrical conductors to connect all collectors on a circuit to the appropriate collection point.

- The utilization of the appropriately sized grounding conductor to connect all the collectors on a branch and at every branch level to appropriate collection point.

- The utilization of the appropriately sized insulated electrical conductor and accessories to connect dc power to charge controllers, inverters and dc panels.

- The utilization of the appropriately sized grounding conductors and accessories to connect to the inverter to the AC panel.

- The utilization of the appropriately sized conductors and accessories to connect the outlet side of the inverter to the AC panel and to the electrical grid.

- The utilization of an appropriately sized lightening conductor system to properly ground the solar tree.

- The utilization of the appropriately sized AC and DC disconnect switches, AC and DC panels, AC transformers, DC-DC converters and accessories.

Battery Energy Storage System for the NATARCH Solar Tree

Battery energy storage is an essential part of every solar power generating system and the design of any such system must be based on the following:

- The utilization of the best and most appropriate battery technology to meet the needs of individual consumers or utility suppliers utilizing the NATARCH solar tree system.
- The utilization of an appropriately sized battery for each NATARCH solar tree or the utilization of a central battery energy storage system for larger "orchard" of NATARCH solar trees.
- The utilization of the best available technology for the creation of the storage location, the safety and protection of the battery energy storage system and accessories.

Cleaning System for the NATARCH Solar Tree

All solar systems are impacted by dust in the atmosphere and provisions must be made to regularly remove as much of the deposited dust from the leaves of the NATARCH solar tree and the design shall be based upon the following:

- The utilization of the best available water spray technology, inclusive of pump, piping, spray heads, valves, electrical systems, control systems, fixing supports and a water storage system, where necessary.
- The utilization of the best available structural design technology to ensure the proper location of water storage tank.

Access System for the NATARCH Solar Tree

The structure of the solar tree will be fairly tall and, on some designs the NATARCH solar trees will be made up in many sections which will require the utilization of appropriate technology to install, maintain and uninstall the complete system and the design for this system shall be based upon the following:

- The utilization of appropriately designed and structurally sound ladders and platforms systems.
- The utilization of material of adequate strength as determined by international standards.
- The utilization of material that is corrosion resistant

- The utilization of necessary insulation systems where appropriate to prevent electrical shock to people utilizing the access system.
- The utilization of an appropriately designed connection system that will facilitate the attachment of the access ladder and platforms to the structural support system of the NATARCH solar tree.

Electrical Output System for the NATARCH Solar Tree

The electrical energy created by the solar tree must be withdrawn from the solar tree if it is to be productively utilized by individuals, medium size businesses, large companies or a utility service and the system must be design according to the electrical codes of all authorities having jurisdiction, the NEC, IEEE and the IBC and may include the following:

- The utilization of an appropriately sized boundary protection systems to prevent people and natural fauna from encroaching on the highly charged equipment, wire and cables and accessories for the transmission substation or other outlet system.
- The utilization of appropriately sized switches, cables, grounding systems, panels and breakers, transformers and conduiting to meet the requirements of the system.
- The utilization of the best available electrical technology that adheres to the requirements of the local authorities with jurisdiction and all appropriate electrical codes.
- The utilization of the best available technology, appropriate methods and practices in electrical power take-off.
- The utilization of the best available lightening protection technology, electrical protection systems and switches and safety systems.

NATARCH Solar Tree Framework

The most important considerations for the NATARCH solar tree framework are the geometric shape of the crown, overall height and the height of the operational system above ground level.

Geometric Shape of The Crown

The shape of the crown of the proposed NATARCH solar tree is critical as it is this shape that will determine the amount of incident solar radiation that each crown can collect on an hourly basis and the shape of the

NATARCH solar tree shall be derived from the typical shape of trees found in nature some of which includes the following shapes pyramidal, cylindrical, ovoid, half ellipsoid and spherical. The importance of each shape, in the field of solar energy, is a function of the projected surface area, the area that the sun sees at each hour of daylight, as this projected surface area will determine the number of solar cells that will be required for capturing and converting incident solar radiation to electrical energy each hour of daylight. In nature, the shape of the trees are a function of their origin and the evolutionary changes that were required to make in order for them to survive as a specie, however, for use in the solar field the shape to be chosen will be determined by surface area only, since the most important criteria is the amount of electrical energy that such trees can produce. In order to make the optimum selection, a comparison of a few natural crown shapes shall be made and the comparison process shall be completed by calculating the surface area of a few of the selected crowns and the selected group for doing necessary calculations shall include the pyramidal, ovoid, and spherical crowns.

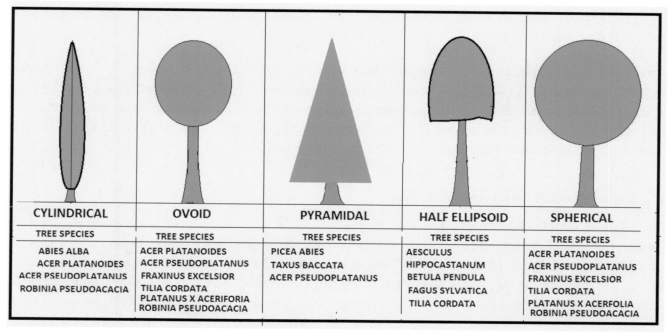

Fig 10.4 Typical Tree Crown Shapes. Source: Mdpi

The shape of the crown chosen for comparison represents a wide cross section of the trees found all around the world and a few of these tree species are highlighted on the diagram

Diameter & Height(ft)	Conical (sq. ft)	Ovoid (sq. ft.)	Spherical (sq. ft.)
10 x 20	480	157	314
20 x 20	1016	471	1256
30 x 30	2286	942	2826
40 x 40	4064	1570	5024
50 x 50	6350	2355	7850

TABLE 10.1 SURFACE AREAS FOR TYPICAL SHAPES

The calculation for each shape was made using the following equations:

- Cone Surface Area, $A = 3.14 \times r [r + \text{square root} (r^2 + h^2)]$
- Ovoid Surface Area, $A = 3.14 \times a/2 \times b/2$, where a = major axis/diameter, b = minor axis/diameter
- Sphere Surface Area, $A = 4 \times 3.14 \, r^2$

An analysis of the areas calculated indicates that the largest surface area is given by the spherically shaped crown at most dimensions and the spherical shaped crown will also present the largest projected area at almost every hour of daylight and based on these metrics the sphere-shaped crown would be selected as the most suitable for the optimal production of electrical energy, however all shapes may be utilized as the solar trees may also be utilized for decorative and aesthetic purposes.

Height of the NATARCH Solar Tree

The height of the NATARCH solar tree shall be based on the height of standard trees and not on either the giants or midgets of the natural forests of the world, however, the height selected here is not an indication of the optimum height of the proposed NATARCH solar trees, as significant structural and energy calculations would be required to determine the most productive, cost effective and safe height for these electrical energy generating trees, calculations which will not be included at this stage. The height of the NATARCH solar tree becomes critical as it must be built to withstand all natural weather events, especially wind events which may have the capacity to topple top heavy structures and the NATARCH solar tree may be considered top heavy depending on the size of the, the area of resistant presented to the wind and eventual structural design that is chosen. Some of the other height considerations that could impact the structural stability of the solar tree would include the overall height and the surface area of the solar radiation collecting component that would be located at the higher levels. Very tall trees with large open or enclosed surface areas could make the structures more susceptible to failure from high winds such as may occur during a hurricane, unless they were built extremely strong at costs that could make these trees uneconomical to build and operate. The average heights of most trees fall into the range of 40-100 feet tall and the height selected for utilization with the NATARCH solar tree shall fall into the middle of this range and would be approximately 70 feet tall or little taller based on the application of a structural system that could be economical to provide and safe to use.

Historically, many mature trees are much taller than this chosen height of 70 feet and trees of many species perform the processes of photosynthesis, collecting solar radiation, collecting carbon dioxide, water and nutrients and processing the same significantly well at similar heights and to heights as high as 500 feet for the tallest tree species. The selection of this height of 70 feet was a preliminary selection only and it is very possible that a NATARCH solar tree could perform very well at much greater heights. It should be noted that any selected height will impact the size of any large-scale installation as the collector field must be designed to prevent the shading of tree by tree during all hours of daylight and taller trees would probably require greater spacing between trees to minimize the impact of shading. The taller NATARCH solar trees could also provide greater benefits if they are widely spaced as this could allow greater usage of the available space beneath the canopy.

Height of the Crown Above Ground Level

The purpose of raising the crown to a significant level above ground is to preserve the land beneath the canopy for utilization in other areas of human activities inclusive of agriculture, crops and livestock, recreation, education and healthcare, however, this height above ground cannot be unlimited as consideration must be given to the overall height of the tree, the height of the required solar radiation collection surface area, the related safety issues and the cost to construct and secure such system at the greater heights. To optimize the overall height of the crown above the ground, this height selected could be considered to be the maximum height of the crops to be planted beneath the canopy or the height of any transportation or land preparation equipment that may be required in the field, this height could be in the region of 15 -20 feet (5- 6 meters) with a maximum of 20 feet or 6 meters. The height above ground could also be higher to favour more under canopy activities but the priority consideration must always be the solar collecting surface area, given the selected height of 70 feet.

Other Considerations

The selected height of 70 feet above the ground and the selected 15 feet clearance of the crown above the ground level would limit the maximum height of the spherical crown section to 70-15 which is 55 feet with an approximate available surface area of 8000 plus square feet on the external surface of the sphere. The combination of the 15 feet height above the ground, the 55 feet height and width of the sphere and any necessary underground support systems could set the overall frame of the solar tree within a, 80 feet by 80 feet cube. If the natural architecture of the tree is followed the roots would spread to two and a half times the width of the crown, however, with proper structural design this large area will not be required. The structurally designed roots of the NATARCH solar tree may not be able to provide the same soil support as the natural roots of a tree would, unless additional measures are taken to protect the soil. The size of above-mentioned containment area would be increased if there is greater spacing between the NATARCH solar trees when the need to minimize shading is added, but this spacing between trees would not add to the structural requirements of the NATARCH solar tree frame.

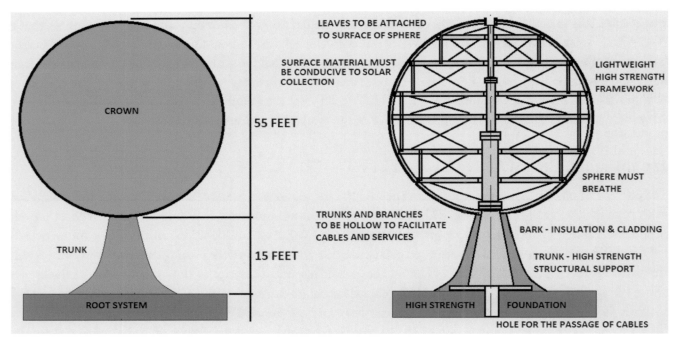

Fig 10.5 Probable General Framework For A Natarch Solar Tree

The figures above highlight the possible basic framework, shape and structure, that could be utilized for the NATARCH spherical solar tree and the size of the sphere, structure and foundation could be adjusted to meet desired electrical output and to ensure that the system is structurally sound, balanced and suitable for each location.

Collectors for The NATARCH Solar Tree

The leaves on a standard tree represent the solar radiation collectors and converters and on the NATARCH solar tree the leaves or the collectors shall serve the same functions but they shall be from manmade material with the necessary properties of photoconductivity, large persistent photoconductivity (LPPC), transparency, high flexibility, high strength and selected from the best of the third-generation solar cell technologies. The design of these collectors or leaves must, however, start with the very basics, the size and shape to set the stage for optimum collection of the incident direct and defused solar radiation.

Leaf Size for the NATARCH Solar Tree

According to Yates et al (2010) "Leaf size and shape influence a range of important physiological processes including photosynthesis, transpiration and thermoregulation" and this indicates that in the natural forests and among plants in general, the size and shape of each leaf is very important for the collection and conversion of solar radiation, the collection and processing of carbon dioxide, water and nutrients by photosynthesis, for the production of carbohydrates and other essential plant foods. The importance of leaf size and shape to natural trees is also an indicator that size and shape of its collectors will also influence the capacity of the NATARCH solar tree to capture and convert solar radiation into electrical energy. The impact of leaf size and shape on collection and conversion shall be one of the determining factors in designing and selecting the optimum size of the collectors, leaves, to be utilized on the manmade NATARCH solar tree. In nature the size of leaves of each species is determined by several factors inclusive of the region of globe, tropical or temperate climate, the local temperatures and the availability of water, which means that in arid and cold regions low levels of water the leaves on the trees will tend to be smaller than those in warm and wet regions, but on NATARCH solar trees the determining factors shall be the ability to collect the optimum amount of solar energy, the ability to efficiently convert the collected solar radiation into electrical energy, how effectively these leaves will fit on the surfaces of the spherical tree and how easy the it will be to mass produce the new solar cells.

The shape of leaves evolved by the many different species of plants over time is generally very important as the shape helps in the collection and utilization of carbon dioxide in the process of photosynthesis, however, since the solar tree will not be performing photosynthesis, a particular shape will not be as important and the shape criteria may be eliminated as an important factor in determining the final shape of a leaf for the proposed NATARCH solar tree. The other dimension that may be of great importance is the thickness of the standard leaf, with the thickness of a standard natural leaf varying from 100- 300 nano-meters depending on the species, as this thickness significantly impacts the process of photosynthesis in a natural plant, as thickness determines the distance the photon and electron must travel during the process. Fortunately, thickness while important for natural trees, may not be as important a consideration for the design of the new collectors for the solar tree as the third-generation solar technologies were already chosen and these technologies are usually much thinner than natural leaves as they are usually built with thickness at the lower nanoscale level, much thinner than leaves.

The surface area of the leaf alone shall therefore be the determining factor in the selection of the leaf size, which must fall into the category of large leaves as produced by broadleaf tree species that produce leaves with surface areas in square inches or large leaves as may be produce by many tropical plants such as the giant taro plant with areas of many square feet.

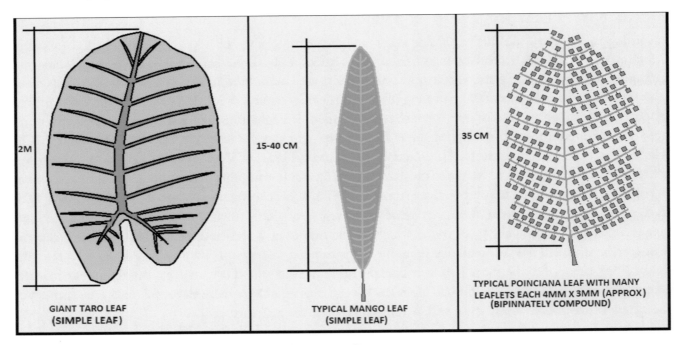

Fig 10.6 Three Leaf Types Highlighting Different Leaf Areas.

The giant taro which was first domesticated in the Philippines island is a perennial herb which was said to have evolved beneath the canopies of forests in the tropical regions of Asia and it is also said that the plant developed its very large leaves, the largest of its type in the world, in order to collect the weak sunlight available beneath the forest canopies.

The mango tree, Mangifera Indica, is a tree that can grow to heights of approximately 30 meters was first domesticated in India and evolved in the India-Burma region of South East Asia with leaves that are relatively small compared to those of the giant taro, the size of the mango tree leaves clearly highlights the difference between a plant that gets direct sunlight and one that receives mostly diffused sunlight.

The poinciana tree, Delonix regia, is a fast-growing evergreen tree that can grow to heights of approximately 10-15 meters, is a Madagascar native that has bipinnately compound leaves which were evolved to present the least surface area to minimize water losses, these trees evolved in hot dry climates.

The average leaf area of the giant taro is generally over 1 meter squared, the mango approx. 0. 016 meter squared and the poinciana approx. 0.0072 meter squared. The average solar modules have areas that are generally 1.6-1.8 meters square, similar to the giant taro leaf, which fits very well with the current presentation of solar systems to incident solar radiation, rectangular flat surfaces tilted at an angle. The surface of a sphere, however, would probably require different sized collectors to fit onto its surface, sizes in a similar range to the size of the mango leaves or much larger such as in the range of 0.02-0.04 meter squared and should probably be a square shape. The size of each solar cell and the method of affixation shall also impact the size of each collector.

Collector Technology for the NATARCH Solar Tree

The decision on the choice of collector technology will require sifting through several current unproven third-generation technologies as there are quite a few such technologies available to choose from inclusive of the Dye Sensitized Solar Cell (DSSC), Perovskite Solar Cell (PSC), Organic PV Solar Cell (OPV), Organic Tandem Solar Cell (OTSC), Quantum Dots Solar Cell (QDSC) and combinations of quantum dots, organic solar cells and other similar technologies. All of these unproven technologies will be reviewed with a view to eliminating the less favourable ones, because even though all of the third-generation technologies are displaying great properties some have less favourable qualities that would not fit with the concept of fitting in with nature and this incompatibility with nature would almost automatically eliminate them, especially those technologies that utilize toxic metals. The ones to be eliminated earliest will includes the DSSC and the Perovskite Solar Cell, the DSSC because of the problems associated with its liquid electrolyte content and the probability of leakage and the Perovskites due to their lead content, their current lack of stability and high failure rates. The selection process shall therefore be focussed on the OPVs, OSCs and OTSCs, QDSCs and the combination of OPVs and QDSCs.

OPV. Organic photovoltaic (OPV) cells consist of two technologies Organic Solar Cells (OSCs) and Organic Tandem Solar Cells (OTSCs) two technologies which were both designed to utilize organic materials, organic plastics, which have the desired photosensitivity and photoconductance properties to produce the photovoltaic effect and these organic plastics are also abundant and readily available for use in solar power generation system with the objective being that the utilization of these materials would make the production of electrical energy from solar radiation relatively inexpensive compared to the earlier solar technologies. These low-cost organic materials also have many other positive desirable qualities inclusive of the following:

- They have the capacity to produce the photovoltaic effect.
- Low production cost combined with high production volumes could ensure that organic PV solar cells become much less expensive to produce than silicon solar cells.
- Lightweight and the high flexibility of organic molecules.
- High optical absorption coefficient which allows a small area of material to absorb a large volume of solar radiation.
- Early projections indicated that it would be a cost-effective technology in the field of photovoltaics, based on current cost.
- These cells required less energy to produce and is therefore environmentally friendlier to produce than silicon cells.
- These solar cells were considered very safe for use as it had no established negative environmental impact.
- Its lightweight and flexibility make it suitable for making things like bags and clothing.
- The material possesses an intrinsic property that facilitates long persistent photoconductance (LPPC).

With all of the very important qualities highlighted above, the OPVs looked like the ideal PV cells that should be utilized in the design and construction of a solar system that is intended to mimic nature in the form of the NATARCH solar tree. Unfortunately, there are some significant drawbacks as OPV cells have been shown to be limited in efficiency, have low stability and low strength relative to the existing solar technologies. However, if low, limited efficiency of 8.3 %, was the only problem it would be possible to select the OPV without question as several thin film technologies currently operate at this level of efficiency, however, it has been established that OPV solar cells currently lack stability and strength and this lack of stability and strength means that anything built using the OPV cells technology will not have a long useful life and therefore could not provide any long-term effective energy or environmental benefit. These negative issues could combine to take the OPV cells out of consideration for use with the solar tree, as while close to the ideal solution, their current faults make them unsuitable. However, based on the current status of the other third generation technologies and the continuing research being done with OPV cells there is a very strong possibility that significant improvements will be achieved in the near future and this means that the OPV technology must be given further consideration.

OTSC. The Organic Tandem Solar Cell (OTSC) which is directly related to the OSC cells was designed and built to overcome the 32 % conversion efficiency limit associated with most solar energy systems and this efficiency limitation was thought to be associated with known losses in the operation of organic PV system. It was established that with the elimination of these two known sources of energy losses, higher efficiency could be achieved in single-junction OSC solar cells and the desired endpoint could be achieved by eliminating the two known major losses, the transmission losses and the thermalization losses. The OTSC design which utilizes two OSC cells connected in series, tandem, successfully overcame these losses, however, while solving these problems, this technology created new problems which must also be eliminated. The creation of tandem cells requires the use of another material, an interconnecting layer, to connect the two cells forming the tandem and this interconnecting layer (ICL) causes several problems inclusive of low intensity conditions in the secondary cell, consequently leading to low efficiency and difficulties associated with mass producing the OTSC cells. The creation of a highly transparent material with adhesive qualities to replace the existing ICL could make the OTSC more efficient and stabile and make it an ideal, cost-effective solution for the NARARCH solar tree. However, until that cost-effective solution is found the OTSC, while a positive step forward with the reduction or elimination of the transmission and thermal energy losses, is not currently a good candidate for selection and must be overlooked for the time being.

Hybrid OTSC. Recent work by Gu et al (2022) has indicated that a hybrid OTSC consisting of a perovskite as top cell and an OSC as sub cell with an improved interconnecting layer can achieve a power conversion efficiency as high as 20.6 % which is similar to or better than the existing generations of silicon solar cell technologies. While this result is a vast improvement on previous OPV systems it was achieved by utilizing a perovskite which contains a large component of lead and lead with it many environmental negatives cannot be utilized on a NATARCH solar tree, whose design is to enhance the environment while producing needed electrical energy. The technology however, proves that a hybrid OTSC could produce greater results than the standard OTSCs and the possibility that the replacement of the perovskite with an environmentally inert QD such as carbon or graphene could make things quite interesting and give new life to this hybrid OTSC.

QDSC. The quantum dots solar cells (QDSCs) in any one of several configurations inclusive of quantum dot-sensitized nano-crystalline titanium dioxide solar cells, the quantum dots dispersed in organic semiconductors polymer matrices cells, quantum dots combined with other technologies such as Perovskites, utilizing any one of the many forms of quantum dots made from different materials or alloys of materials and utilizing any one of two forms of QDs, core and core and shell, brings QDSCs to the fore as major contenders with many significant possibilities. While all of these systems have the promise of producing higher conversion efficiency outputs at later stages of development, some developments have already been found to have certain problems that could lead to their early elimination from the selection process for use on the NATARCH solar tree. Some quantum dots are made from material that are known to be toxic and very harmful to humans and the environment, however, it has also been established that these toxic materials make some of the most efficient solar cells, materials such as lead sulphide are known to be toxicants and it also makes solar cells with the highest conversion efficiencies. The QDSC systems that are early contenders for elimination would include the quantum dot-sensitized nano-crystalline titanium dioxide solar cells, QDs made from lead, lead compound and lead alloys and QDs made from other materials which have toxic properties that could negatively impact both humans and the natural environment. The quantum dot-sensitized nano-crystalline titanium dioxide solar cells have been shown to have a low energy conversion efficiency of 1.7 percent and they have also been shown to be not very reliable. This combination, low conversion efficiency and low reliability would eliminate this configuration from long-term usage, in its current stage of development. The QDs made from lead and other toxic material can be automatically eliminated due to their negative impacts upon the health of humans and the environment, even though lead and other toxic metals such as cadmium have produced some of the most efficient QD solar cells and QD configurations.

The best contenders among the current crop of QD technologies are the technologies which have produced the highest conversion efficiencies to date and which do not contain lead, the worst environmental offender. Unfortunately, it is very hard to find a more efficient QD than those which contain lead as a major component as the latest reports on research in QD technology from a laboratory at the University of Queensland in Australia will show. These reports indicated that this laboratory has produced a QD, made from a Perovskite material, which has a conversion efficiency of 16.6%, the highest efficiency recorded in the world from QDs. This reported result, which has been confirmed by NREL, is a twenty five percent increase over the previous highest recorded efficiency of 13.4 % achieved in 2017 by NREL laboratories, while using a QD made from a similar Perovskite material. Unfortunately, the most efficient Perovskites all contain lead as a major component and raises the very important issue of negative environmental impacts, which could be mitigated if another group IV or any other material, which is less toxic, could be substituted for lead in the Perovskite crystalline structure and produce the same high efficiency. A report by Zhang et al (2018) has shown that significant effort has been made to find a substitute to replace lead in the Perovskite crystal and this report indicates that many metals have been used to successfully relace lead but none of these metals have to date produced the results obtained by lead in Perovskites for PV systems. Several metals inclusive of tin produced significant results, however, much more work would need to be done before it could be used to successfully replace lead. This inclusion of lead in the Perovskite would significantly reduce the chances of this highly efficient QD solar cell being selected for the NATARCH solar tree, however, finding an equally efficient substitute for lead would make it the ideal collector material.

The selection process at this point could be considered at an end and the OPV with all of its present faults selected as the default choice, however, there is still one QD that should be reviewed and given consideration even though it does not rank as the QD with highest efficiency, but it is ranked as being biocompatible which is much better than all of the other technologies, this is the carbon QD, which include both carbon and graphene, another form of elemental carbon.

Carbon and Graphene Quantum Dots (CQDs & GQDs)

Carbon is the foundation of all life on planet earth and has been central in the utilization of solar energy to foster the growth and development of all plant and animal species that today calls planet Earth home and it has also been the central source of indirect solar energy, with the exception of nuclear energy, that humans have utilized since the beginning of human existence. It is therefore very interesting that carbon, in two forms, could possibly be the foundation upon which to build a new generation of energy systems that will utilize more efficiently the energy of the sun to eliminate the excessive, harmful presence of carbon dioxide in the atmosphere, with these two forms of carbons and the sun, forming one of the longest natural partnerships on record. According to Paulo et al (2016) the most important properties of the carbon and graphene, which make them very good candidates for use as quantum dots for photovoltaic systems are the fact that they are abundant, inert, non-toxic, biocompatible, can be made from biomass, and have great light harvesting and conducting properties. According to Lim et al (2014) "Carbon Quantum Dots, CQDs, are typically quasi-spherical nanoparticles comprising amorphous to nanocrystalline cores with predominantly graphitic or turbostratic carbon (sp^2 carbon) or graphene and graphene oxide sheets fused by diamond-like sp^3 hybridised carbon insertions" Lim et al (2014) also indicated that these carbon nanoparticle have the potential to greatly improve many areas of medicine specifically in the areas of bioimaging, drug delivery, chemical sensing, biosensing and several other related areas of medicine, due mostly to the inert chemical make-up, tuneable fluorescence emissions, excellent physicochemical and photochemical stability and the ease with which they can be manipulated to produce the desired CQD configuration. The Lim et al findings highlights mostly the use of carbon quantum dots in medicine and other related areas but not in photovoltaics in particular, as they considered that the quantum yield of CQDs were too low for PV use, however, significant improvements have taken place since 2014, that have greatly improved the value of CQDs to solar energy with Vercelli (2021) ascribing the following properties to the updated CQDs, non-toxic, easy synthesis, water soluble, chemical inertness, wide range light absorption, solar energy conversion, fluorescence emissions, photocatalytic properties, up-conversion fluorescence, good charge transfer and separation ability and high quantum yields, all of which could make CQDs equivalent to inorganic semiconductor quantum dots and very suitable for use in photovoltaic energy systems.

The research reports with respect to graphene quantum dots (GQDs) states that the impact of carbon may be even greater as according to Vibhute and Shukla (2016) graphene is a new super material in the field of solar energy conversion with properties such as high thermal conductivity, high electrical conductivity, high transparency, great mechanical strength, great flexibility, high aspect ratio, large specific surface area and it is the thinnest and strongest material ever measured. Graphene also has other properties which makes it excellent for solar energy conversion systems inclusive of:

- Electron and hole charge carriers display immense natural mobility
- It has a tuneable band gap
- It has the lowest known operative mass of any material, which is zero
- It can travel relatively very short distances, micro-meters, without diffusion at ambient temperatures.
- It has the capacity to maintain current densities at six orders higher than copper
- It displays record thermal conductivity and rigidity
- It cannot be penetrated by gases

These properties coupled with the fact that the material required to manufacture CQDs are in great abundance, are readily available at a low-costs, are environmentally friendly and the cost to manufacture is relatively inexpensive, makes CQDs an attractive alternative for use to produce electrical energy from the sun. The research completed to date, however, has shown that CQDs have the capacity to improve the output of several OPV systems inclusive of the organic solar cells (OSCs), the dye sensitized solar cells (DSSC) and perovskite solar cells when utilized in one of several capacities inclusive of, as electrocatalysts, sensitizers, electron acceptors, electron co-acceptors, electron donors, electron blocker, hole extraction layers, electron transport layers, charge carrier layers, dopant, counter electrodes and as energy-down-shift material which greatly improves the range of the spectrum that will impact the cell, UV to NIR. The CQDs can also be used to greatly improve the output of inorganic semiconductors nanoparticles such as Zinc Oxide and the operations of the bifacial DSSC cells. Unfortunately, to date no research reports have yet indicated that CQDs cells would make great solar radiation collectors or converters on their own even with the many significant properties that they display inclusive of high fluorescent emissions, wide range light absorption, solar energy conversion, good charge transfer and separation ability and high quantum yields. In fact, the best portrayal of CQDs is that of a great enhancer and facilitator for other photovoltaic systems inclusive of the DSSC, the OSCs, perovskite solar cells and other PV systems. The benefits provided to these OPV systems includes reduced electron-hole pair recombination, increased charge density, and boosted electron mobility, all of which combine to improve overall output and efficiency of the OPV cell, Kumar et al (2022).

The final choice of collector material to be selected for use on the NATARCH Solar Tree, must be from this hybrid group, however, based on the prior assessment of the perovskite and DSSC solar cells weaknesses, the DSSC and PSC in combination with carbon quantum dots (CQDs) while having the possibility of producing great results and great promise for the future, would be eliminated from selection at this time due to their established problems and the focus is now placed on organic solar cell (OSC) and CQD combinations. It should be noted that according to Vercelli (2021) the conversion efficiencies of the most up to date OSCs and OTSCs have risen greatly without the input of CQDs, to levels of 17% and 14.2 % respectively and this reality suggest that the incorporation of the CQDs could increase the conversion efficiency values to significantly higher levels.

OSC and CQD

The combination of OSCs and CQDs has to date created significant results in many ways inclusive of combing to give the first all-weather PV systems, a system that produces electricity with bright sunlight and when the rain is falling and this capacity, all weather, is a function of the property graphene, a Carbon Quantum Particle, the chemistry of rainwater and the reaction between them both. Rain water typically contains salts which separate into positively and negatively charged ions of sodium, calcium and ammonium and the graphene surface has the capacity to bind with these ions and when the rain falls these positively charged ions come into contact with the graphene contained in the solar cells and the rainwater becomes rich in positively charged ions and negatively charged detached electrons creating a pseudo-capacitor which has the capacity to produce voltage and current which allows the all-weather solar cells to produce electrical energy on both sunny and rainy days.

These GQDs also display ultralong persistent emissions that can last for as long as one hour and according to Tang et al (2017) these all-weather GQDs can produce power conversion efficiency as high as 15.1% in dark condition. In addition, these GQDs have been shown to improve power conversion efficiencies in most OPV systems inclusive of DSSC 13%, a GQD/silicon hybrid 14.5% and the GQD perovskite hybrid 15%.

Based on all of the above the selected collector technology shall be an OSC GQD hybrid with an efficiency above 13 % or the current best in class. The solar cell shall also be transparent, have a high long persistent photoconductance, shall designed as a tandem cell and have two faces as an option on the open NATARCH solar tree designs.

Alternative Collector Design.

The design of the collector technology as proposed above was completely based upon the use of third generation proven and unproven young technologies with an eye to the future and on the reduction or elimination of the environmental costs associated with the mining and production of the most efficient PV cells to date, silicon wafers, and the elimination of second-generation thin film solar cells that contain toxic material that may be harmful to humans and the environment alike. However, if these proven and established first and second generations solar technologies could be modified to fit into the surfaces of the proposed spherical NATARCH solar tree, as in BIPV systems, the older technologies may also be utilized to make the NATARCH solar tree more quickly available in areas of the world where it could provide benefits that are greatly needed . Basically, the older technologies could be used in the solar collectors of the NATARCH solar tree while the development of the third-generation technologies is completed, after which the use of the first and second generation technologies should be discontinued to begin the utilization of the newer technology on new NATARCH solar trees with the maturing of the more efficient and more environmentally friendly third generation technologies.

STRUCTURAL SUPPORTS FOR THE NATARCH SOLAR TREE

The structural supports for a natural tree are their roots, trunk, branches, limbs and twigs which give support to the leaves and the fruits of the tree and the NATARCH solar tree shall require similar supports for its collectors and accessories that will be producing the electrical energy from solar radiation. The NATARCH solar tree should be structurally similar to a natural tree with trunk, branches, limbs and twigs to ensure the optimum presentation of all of the solar collectors to the sun at each hour of the available daylight. The most important requirements for the structural system of this NATARCH solar tree shall be the material of construction, the structural integrity of all of the major elements separately and the structural integrity of all elements combined to form the frame and foundation of the NATARCH solar tree. The major elements consist of foundation, trunk(column), branches(beams) and limbs (small beams).

Foundation

The foundation shall consist of heavyweight, high strength, steel reinforced, environmentally friendly concrete that will be buried beneath the soil, with a span and volume that is adequate to firmly support the weight of the crown, hold the frame of the solar tree firmly anchored to the ground and to act as a counterbalance to all forces that may be exerted upon the crown, especially wind and snow. The foundation shall also have a hole, 1.0 meter in diameter or as recommended by the electrical designer and the structural designer, to provide access for all conduits and cables going to and leaving the NATARCH solar tree.

To mimic the role of the tree roots as a natural soil conservator, natural biodegradable material could be attached to the foundation and be spread out to two and a half time the crown diameter at levels below the surface, this material along with any grass that may be planted on the surface would hold the soil together and prevent erosion by water and wind.

Columns

The crown shall be supported by a high strength hollow section, preferably circular, which shall act as a column or columns as may be required, to support and balance the load of the crown. If more than one column is utilized then both have to be spaced to adequately support the load and maintain a balance during events like very high winds. The column shall be constructed from high strength carbon fibre, a material which is five times stronger than steel and twice as stiff or a material of similar strength and durability, have high resistance to abrasion, high corrosion resistance, high tensile, compressive and torsional strengths and be cost effective to use. The material must facilitate all standard mechanical joining methods, as may be required, to join sections together or whatever new method of fabrication and joining is relevant. The hollow section will give access to all necessary conduits and cables required, and shall also provide access to the branches that will be supported by and are attached to it. The surface of this material must also be amenable to the attachment of a protective skin with a rough outer surface by utilizing a high strength adhesive.

This column shall be attached to the foundation by a high strength flange, nuts and bolts and or other acceptable structural joining methods as designed by a structural engineer and approved by the relevant authority have jurisdiction. The size of the column shall be reduced with the increasing height of the tree and shall terminate in a structurally sound manner at the top of the tree, to prevent the collection and accumulation of rainwater inside the hollow of the column. Access panels must be provided at the lower end of the column or columns and at all branch levels to provide necessary access to wires, switches, circuit breakers or fuses that may be located on the inside of the column.

Branches

The branches shall be attached to the trunk of the tree starting at the lowest level that will ensure that the bottom of the branch is at a minimum of 15 feet above the ground level and these branches continue on up the trunk to the top of the tree with the size of the branches reducing as required at the higher levels. The branches shall be constructed from the same material as the trunk and shall be sized to support the loads of the other structural element which it supports inclusive of the "limbs", "twigs" and the collectors at each level. This way the branches at the middle will be the largest and the those at the top and bottom will the smallest based on the spherical shape of the tree, however on the modified partial spherical design the branches shall be largest at the bottom and smallest at the top. These branches shall also be hollow to ensure access for the necessary wires, cable and conduits that will be carrying control signals and electrical current to and from the branches. The maximum number of major branches at all levels shall be a function of the loads to be carried at each branch level and the strength of the material available for their construction. The number of branch levels shall be designed to meet the specific electrical output requirements of each solar tree and to provide the necessary access to all solar collecting surfaces and fixtures to facilitate necessary maintenance. If a maximum height between levels is established at 6 feet, a 55 feet high crown will have nine branch levels with reducing branch sizes from the bottom branches all the way to the top depending on the design of the particular NATARCH solar tree. The ends of the major branches shall also be constrained to prevent excessive movements especially during a high wind event and the constraints shall be made from the same or similar high strength material as the branches, only strong flexible bands and or cables which shall be applied in two plains, vertically and horizontally.

Limbs

The limbs are smaller versions of the branches with respect to their general make up and as on natural trees they are outgrowths of the branches and have no direct connections, only vascular, connections to the trunk and in the case of the NATARCH solar tree electrical and electronic control connection to the trunk. The limbs support other mini limbs and twigs which carry the leaves and present them to the sun at locations and angles that ensure the optimum collection of solar radiation from the sun. The structure of the limb for the NATARCH solar tree shall be made of similar material as the branches with similar carrying capacities and strength, they shall also be hollow to facilitate the installation of the conduits wires and cable that are required to provide control input and electrical current take off from the collectors. However, the design may not require or utilize the typical "limbs" but good horizontal support between branches that will be supporting the collector surfaces as shown on the original design and alternative design 1.

Collector Supports (Twigs)

The proposed sizes of each collector or leaf, 0.04 meter squared, would require the use of a significant system to safely hold approximately eighteen thousand 0.04-meter square solar cells, in the collector of a 50 ft diameter sphere, in place on the branches and limbs of the proposed NATARCH solar tree and utilizing such an arangement would conform to the idea of using the natural architecture of a trees to create an effective tree solar energy generating plant. However, this would also be creating many thousands points of failure, which no efficient system of energy generation could afford, as each solar cell would require a specific fastening system to get them securely fastened to each limb. The solar cells would preferably be attached to flat surfaces in a manner similar to how solar cells are currently connected to modules, with the modules organized to fit into a circular pattern as shown on diagrams below.

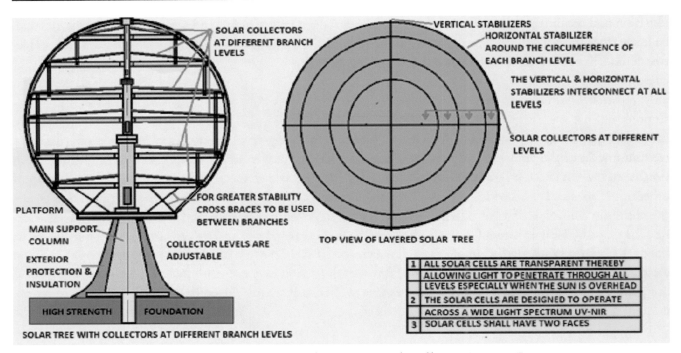

Fig 10.7 Proposed Natarch Solar Tree Structural Support And Collector Layout Design

The above proposed design is based upon the concept of collectors(leaves) being located in a similar pattern as leaves on a natural tree, with the limbs and twigs covering the full length of the branch up to the trunk, depending on the tree species. The design of the lower trunk as shown in this diagram is optional as it serves both as an aesthetic and as a material saving device, which leaves the option for the trunk to be designed as a straight high strength cylinder connecting the foundation and the upper platform. Utilizing an enlarged straight cylinder-shaped trunk would add to the space available for installing AC and DC disconnect switches, electrical switch gear, combiner boxes, battery chargers with MPPT, Battery Energy Storage System and all of the other delicate electronics associated with the safe and efficient operations of a solar power generating system.

The above design in Fig 10.7 is based on the use of the spherical tree shape and therefore the design was confined to the shape presented above even with its obvious disadvantages, however, if the spherical shape is modified the total surface area available for collecting direct solar radiation can be significantly improved as shown in FIG 10.8 below. This design while an improvement on the more complete spherical shape still has some shortcomings as while it allows more areas to get direct solar radiation there will be some areas that will still only get diffused radiation and secondary radiation after it has passed through the collector on the branch above. The proposed collector material and the inclusion of the right electronics inclusive of power optimizers, on this NATARCH solar tree could ensure that diffused radiation and secondary radiation would produce significant amount of electrical energy. This design can also be modified to eliminate all of the areas that would be receiving diffused sunlight only, most of the time, leaving only the areas that will receive direct radiation only, however, eliminating these sections will not increase the total surface areas available or increase the amount of solar radiation that can be collected on an hourly or daily basis. It would however, reduce the cost of this design and the unit cost of each tree but not necessarily improve the overall performance of this design.

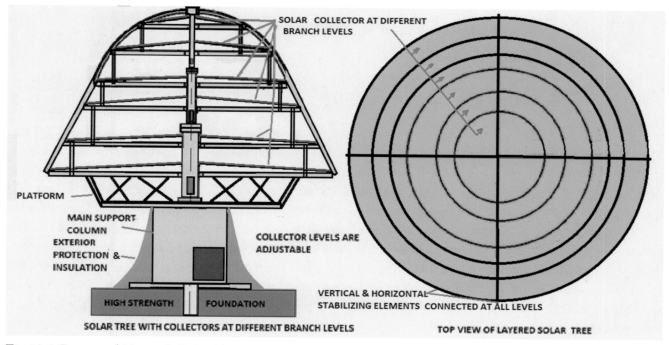

Fig 10.8 Proposed Natarch Solar Tree Structural Support And Collector Layout Design Alternative 1

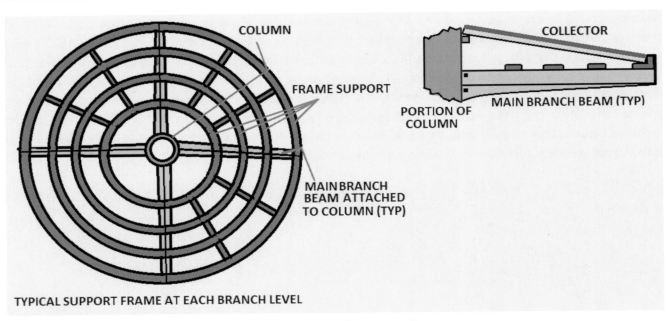

Fig 10.9 Typical Branch Level Structural Frame For Alternative Designs

Another alternative design would utilize the external surface area of the true sphere only, with the material forming the skin around the sphere having the solar cells in-built, as is utilized on BIPV systems. The utilization of the external surface of the sphere only, would also impact the structural requirements of the NATARCH solar tree, as this would eliminate the need for restraining the ends of major branches to prevent excessive movements during wind events, eliminate a main column going up the full height of the tree and could also impact the size of the components used to form the support structures and even how the structure is fabricated and installed on the support column, and these specifications all leads to the use of a system that could meet all the structural requirements as set out earlier, high strength, lightweight, resistant to wind and other natural events, matching a structural system that already exist in the form of a geodesic dome or similar designs. Geodesic dome structural designs have been successfully utilized to build large spherical domes of many types and for many uses all around the world and such a design would simplify the structural design considerations and simplify the construction and installation processes. The inside of the dome would also provide more than adequate space for the installation and servicing of electrical systems, control systems and accessories required for the operation of the solar system and it could also easily facilitate the creation of necessary access points on the dome that would support the work of the maintenance team. The major advantages of the structural design of the geodesic dome according to Hawkins et al, is that due to their spherical shape they are highly resistant to wind and snow loads and has exhibited the capacity to resist even hurricane strength winds, they are energy efficient, require less building material, are relatively inexpensive to construct, has an even distribution of structural loads thereby making the systems very strong and they are generally lightweight and very sturdy.

The utilization of the surface area of the sphere only for the collection of solar radiation, would also eliminate the need for the consideration of the other natural structural elements such as branches, limbs and twigs in the form in which they exist in nature on trees or as they would exist on the other versions of the NATARCH solar tree and the elimination of these many elements and the move towards a more simplified structure provides an additional impetus to utilize the more sophisticated geodesic dome design which could reduce the cost of the utilization of the external surface areas of the spheres, especially if the surface area of the spherical dome is greater that the surface area achieved by the other two designs presented. One other possible advantage of utilizing the external surface area of the spherical dome design is that the surface of the collector would probably be less complicated to clean than the designs with many branches.

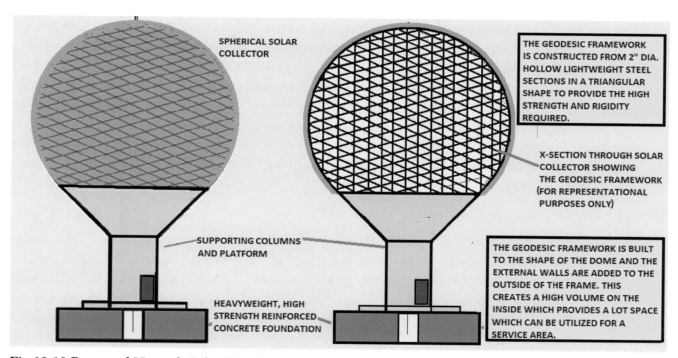

Fig 10.10 Proposed Natarch Solar Tree Structural Support And Collector Layout Alternative 2

A review of these three designs will indicate that all three have strong points, however, the first two designs have significant shading problems that may impair their ability to collect the optimum amount of solar radiation and consequently impact their ability to deliver the maximum output of electrical energy per unit cost to build and operate. The first design does not fully expose all of its surface area to direct solar radiation and its structure may be susceptible to a significant weather event. The first alternative exposes much more surface area but it is based on the same structural design that may not be able to withstand any significant weather event and these obvious weak points would points to the selection of second design alternative which exposes greater surface area and has a superior structural design. The final decision on which systems should be utilized, however, should only be made after a detailed analysis of both of the proposed structures and three collector surface layouts.

One other element of these designs that should be given more consideration is the use of the proposed columns under the spherical crowns on all three designs, these columns could be designed and built to serve only as the structural support for the crown and as machine room space, or as a decorative element only and or as an environmental element that would be conducive to the support of microbes, fungi and other small plant life that would be beneficial to the environment. The most beneficial outcome would involve a combination of uses for the trunk similar to the natural use of trunks on trees, which provides structural support, act as an energy storage area and supports a vast ecological system for other life forms, to perform all of these functions the main support column would have to be made larger to accommodate an area for a battery energy storage system, a general utility space and a large textured external surface area to support a microscopic ecological system and added protection from impact by humans and animals. On large utility scale projects, the trunks of the trees should be utilized for structural support purposes mostly as the battery and energy storage requirements would be met by the utilization of a large utility scale battery that would be located in a separate space from the NATARCH solar tree.

Control Systems for the NATARCH Solar Tree

The efficient collection, conversion and the distribution of the electrical energy produced form solar energy systems usually requires the use of several very important electronic control systems inclusive of charge controllers, maximum power point tracking (MPPT) systems, power optimizers, DC-DC converters, DC/AC inverter-controller which contains an adaptive logic control system, an energy management system and a power control unit on the solar system side. In addition to those basic elements the NATARCH solar tree PV systems would be provided with monitoring and control systems that will sense changes in temperature, voltage and current and have the control capacity to react to prevent any of these values getting out of design ranges and causing destructive events. The control system of large commercial and utility scale systems must be linked with a communication system that will report all out of bound elements, a fire control system that will take action to suppress the spread of any ignition and fire event, the internet, weather stations, SCADA and interfacing devices that will aid the grid operator to make necessary adjustments as required. On the output side where the solar systems must interface with residential, commercial, industrial and institutional facilities and then the electrical grid, other electronic control and protection systems must be utilized to ensure the smooth and consistent flow of energy from the solar power generating plant to the end users and the flow of energy from the grid to residential, commercial, industrial and institutional facilities that are tied to and are still dependent upon to the grid. The electronic controls and protection systems on the outlet side should include the smart meters, smart loads, the harmonic filters, voltage and current sensors, frequency sensors, communication systems, electronic switching systems, systems for the overall management of the solar energy production and internet connectivity to ensure effective communications on both input and output sides where necessary. A scaled down version of these same controls systems may also be required and utilized when the NATARCH solar trees are utilized for single-family residential units and small commercial scale operations. This scaled down version would utilize the most advanced grid-tied inverter, charge controller, battery energy storage system and a smart meter. In general, for large commercial and utility scale applications of the NATARCH solar trees the latest technological versions of the systems as proposed by Sandia Laboratories and shown on FIG 5.4 highlighting advanced control systems combined with an advanced utility scale Battery Energy Storage Systems, should be utilized.

Electrical Systems for the NATARCH Solar Tree

All electrical work to be done on these solar trees must follow all relevant electrical codes inclusive of the NEC, NFPA, IBC, the codes of all local jurisdictions and all technology specific codes. A general layout of electrics for a NATARCH solar tree is as shown in Fig 10.11 below and that for a small commercial solar tree installation could take the format as shown in the diagram in FIG 10.12 below. The electrical infrastructure recommended for utilization with the NATARCH solar tree shall include conduiting, electrical wires and cable, DC disconnect switches, AC disconnect switches, combiner boxes, DC/AC inverters, panel boards, fuses and circuit breakers, transformer, combiner panel board and grid connection cables and accessories. Typical panel boards, combiner panel boards and transformers for the larger system as shown in Fig 10.12 below.

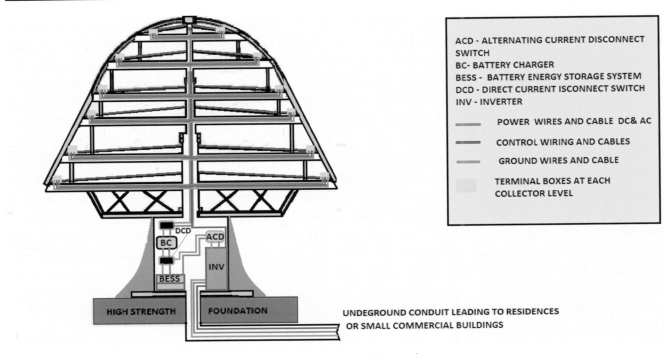

ACD - ALTERNATING CURRENT DISCONNECT SWITCH
BC- BATTERY CHARGER
BESS - BATTERY ENERGY STORAGE SYSTEM
DCD - DIRECT CURRENT ISCONNECT SWITCH
INV - INVERTER

—— POWER WIRES AND CABLE DC& AC

—— CONTROL WIRING AND CABLES

—— GROUND WIRES AND CABLE

TERMINAL BOXES AT EACH COLLECTOR LEVEL

DCD
BC   ACD
INV
BESS
HIGH STRENGTH   FOUNDATION

UNDEGROUND CONDUIT LEADING TO RESIDENCES OR SMALL COMMERCIAL BUILDINGS

Fig 10. 11 Basic Wiring Of A Natarch Solar Tree For Small Instalations

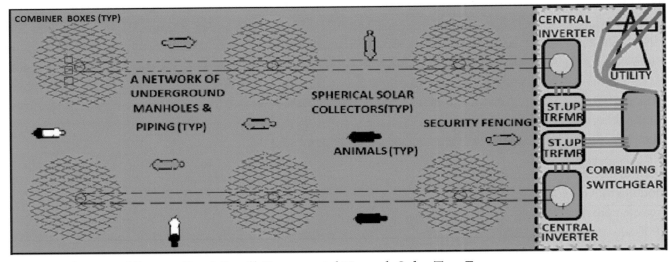

COMBINER BOXES (TYP)
CENTRAL INVERTER
UTILITY
A NETWORK OF UNDERGROUND MANHOLES & PIPING (TYP)
SPHERICAL SOLAR COLLECTORS(TYP)
SECURITY FENCING
ANIMALS (TYP)
ST.UP TRFMR
ST.UP TRFMR
COMBINING SWITCHGEAR
CENTRAL INVERTER

Fig 10.12 Proposed Layout For A Small Commercial Natarch Solar Tree Farm.

The electrical system for a utility scale size project consisting of thousands of NATARCH solar trees would necessarily have to be scaled up to match the electrical loads and must be designed and built to meet all regulations and codes as specified. The same also applies to smaller installations that may include one of two units, the electrical installation must also meet required regulations and codes and must be sized to meet the expected electrical load for these small installations.

Another item of critical concern for the electrical systems on any such mixed-use facility, in this case power generation and agriculture, is the protection of all elements of the solar power generation systems from damage by animals and humans alike and the protection of human, flora and fauna against every possible inadvertent event that may be due directly to the operation of the NATARCH solar tree power plant. The necessary protection will require the provision of the most basic and hardiest protection systems that are available inclusive of a fencing of high strength and durability to create the necessary physical separation to ensure that all unnecessary physical contact between the plant and the surrounding community of people and animals is eliminated, the installation of a properly designed lightening protection system to properly ground all elements of the systems and the utilization of the required electronic switching and control systems to prevent all flows, in or out, that could negatively harm the power plant and or the electrical grid and consequently create negative impacts on the peoples and communities served by the NATARCH solar tree power plant. The basic protection should also include the addition of a "bark" on the external side of the column to reduce any impact that may be caused by animals or humans, this "bark" would be insulating material that would also prevent the transfer of any current to the surface that would harm humans, flora and fauna.

The electrical backbone for these NATARCH solar trees shall be similar to the electrical systems utilized on all solar PV systems and shall consist of appropriately sized conduits, cable and wiring, combiner boxes, AC and DC disconnect switches, switch gear inclusive of panels, breakers and fuses, inverters, LV/MV and MV/HV transformers, LV/MV and MV/HV switch gear, grounding system, battery charge controllers, battery energy storage system and combiner boxes with overcurrent and overvoltage protection switches and monitoring equipment.

The amount of each of the mentioned gear required for the NATARCH solar tree will be a function of the number modules that the total surface area of each tree will be divided into, with a standard module having a total surface area of approximately 14-18 square feet (1.3-1.7 square meters) and a spherical solar collector with a surface area of 8000 square feet will contain approximately 444-571 modules depending on the size of the modules selected and at 20 module per string we are looking at approximately 23-29 strings to be wired and brought to three combiner boxes with accessories, based on ten strings per combiner box. The amount of other support equipment required will also be a function of what type of system the NATARCH solar tree(s) are utilized in, such as small residential system, a commercial system or a utility scale system and whether these systems utilize a battery energy storage system. The number of modules, strings and combiner boxes, as calculated here, will all be multiplied by the number of trees in each system and likewise the number of other electrical components will also be function of the size of the overall NATARCH solar tree energy plant. As the systems get larger and more complicated more components will be required and more care has to be taken to ensure compliance with all electrical regulations and codes. The same consideration, as above, must also be given to all three alternatives of the spherical solar tree with respect to the electrical systems requirements to make each design functional.

Battery Energy Storage System for the NATARCH Solar Tree

The NATARCH solar tree systems will require the utilization of the best and most appropriate battery energy storage system technology available today to ensure that the energy collected and not utilized immediately will not be lost and the base of this technology will be the batteries that make it all possible. There are currently several battery energy storage systems technologies available with a number of batteries made from different materials which are considered the best in the field of battery technology and this group of material includes Lithium, Cadmium, Nickel, Nickel-Cadmium, Sodium-Sulphur, Lead-Acid, advanced Lead-Acid, Zinc-Chloride, Zinc-Bromide, reduction-oxidation flow batteries and a super-capacitor. These materials and technologies combined makes some of the most advanced batteries and battery systems available today and a comparison of a few of these technologies are highlighted in the table below.

TECHNOLOGY	ENERGY DENSITY (kw/kg)	ROUND TRIP EFFICIENCY %	LIFE SPAN (YEARS)	ECO-FRIENDLINESS
Lithium-Ion	150-250	95	10-15	Yes
Sodium-Sulphur	125-150	75-85	10-15	No
Redox Flow	60-80	70-75	5-10	No
Nickel-Cadmium	40-60	60-80	10-15	No
Lead-Acid	30-50	60-70	3-6	No

Table 10.2 Current Leading Battery Technologies and Comparison. Source: Korea Battery Industry Association 2017.

The above table highlights only the top five battery technologies and there are many other such technologies currently available and in use, some of which are as follows:

- Lithium – Several other batteries based on Lithium inclusive of Lithium cobalt oxide, Lithium manganese oxide, Lithium nickel manganese cobalt oxide, Lithium iron phosphate and Lithium titanate,

- Nickel – There are a few batteries based on Nickel inclusive of Nickel- Metal Halide.

- Redox flow – There several other designs of the redox flow batteries inclusive of Vanadium redox battery (VRB), Polysulfide- bromine battery (PSB), Zinc-Bromine (Zn-Br) battery.

There are also several other battery technologies currently under development inclusive of an updated version of Lithium-Ion, Zinc -Air, and Lithium- Sulphur technologies, while research and development activities are also currently in progress on Lithium -Air batteries. The batteries currently utilized in most BESS systems are usually made with Lithium-ion which is the most efficient, durable and only eco-friendly battery technology that has been created to date, as indicated in the Table 10.2 above.

The batteries by themselves, however, do not constitute the entire Battery Energy Storage System as the total system usually consist of the battery pack connected in series and parallel to create the size of the battery required, support racks and several other very important components such as the Battery Management System (BMS), Battery Thermal Management System (B-TMS), Energy Management System (EMS), System Thermal Management System (STMS), System Control and Monitoring (SCM), Voltage and Current Sensing Devices, Temperature Sensing Devices, Supervisory Control and Data Acquisition System (SCADA) and the required power electronics to tie everything together. These Battery Energy Storage Systems have been and are still being built in different sizes to meet the specific needs of the different entities that utilize renewable sources of energy, with sizes ranging from those utilized for single family residential units, commercial and Industrial facilities to the largest sizes utilized for utility scale systems. The current five largest suppliers of BESS systems would include NextEra Energy Resources #1, Toshiba #2, Tesla #3, Sonnen GmbH #4 and General Electric #5 and the five largest BESS systems constructed to date for utility scale power plants to date, are as follows:

- Vistra Energy Corporation - 400 MW/1600MWh, Moss Landing Energy Storage Facility, California

- Florida Power and Light- 409 MW/900 MWh, Manatee Energy Storage Center Project, Florida

- Neoen- 300 MW/450MWh, Victorian Big Battery, Near Geelong Australia

- NextEra Energy Resources -230 MW/920 MWh, McCoy Solar Energy Projects BESS, California

- PG&E- 182.5 MW/730 MWh, Elkhorn Battery, California

Size of NATARCH Solar Tree

The NATARCH solar tree can vary in size based on their surface area and designed electrical output of the tree and each tree or assembly of trees will require BESS systems designed for their particular sizes and no one battery can fit all uses. All NATARCH solar tree PV systems inclusive of residential, commercial, industrial and utility scale types require specific size BESS systems based on their electrical output and mode of operation. These BESS systems can vary in size as follows:

- Residential Systems – 15-100 KW
- Commercial Systems – 0.5- 20 MW
- Industrial Systems – 20- 50MW
- Utility Scale Systems – 100- 500 MW

The existing BESS systems are usually designed with a fire suppression system to suppress any fire that may start due to any excessive increases in temperatures, temperatures that the battery management system, BMS, would have failed to prevent by shutting down the battery operations due to excessive temperatures. This fire suppression system which is usually built into the BESS system, utilizes a suppressant that is suitable for working with electrical systems and usually contain a FM-200 chemical, a non-global warming fire suppressant or equivalent.

All BESS systems utilized with the NATARCH solar trees shall utilize only batteries that contain eco-friendly Lithium-ion technology, a BMS system, an Energy Management System and or a SCADA and a fire suppression system utilizing FM-200 or approved equivalent.

The NATARCH Solar Tree Access System

The NATARCH solar tree is an electromechanical system and like all such systems there will always arise a need to service and do necessary repairs, which requires that a suitable access system be provided that will allow crews to safely access all areas of the system. This access can be provided by a system of access doors, ladders and platforms or catwalks depending on the size of the unit. For the smaller system external access may be provided by the use of scissors lifts or similar equipment while on the much larger systems the stairs and platforms may be built on the inside with direct access to all critical points. All stairs and platform must be constructed from, high strength, corrosion resistant, durable material such as steel or carbon fiber and must be structural sound to support the weight of at least two technicians at the same time.

The Cleaning Systems for the NATARCH Solar Tree

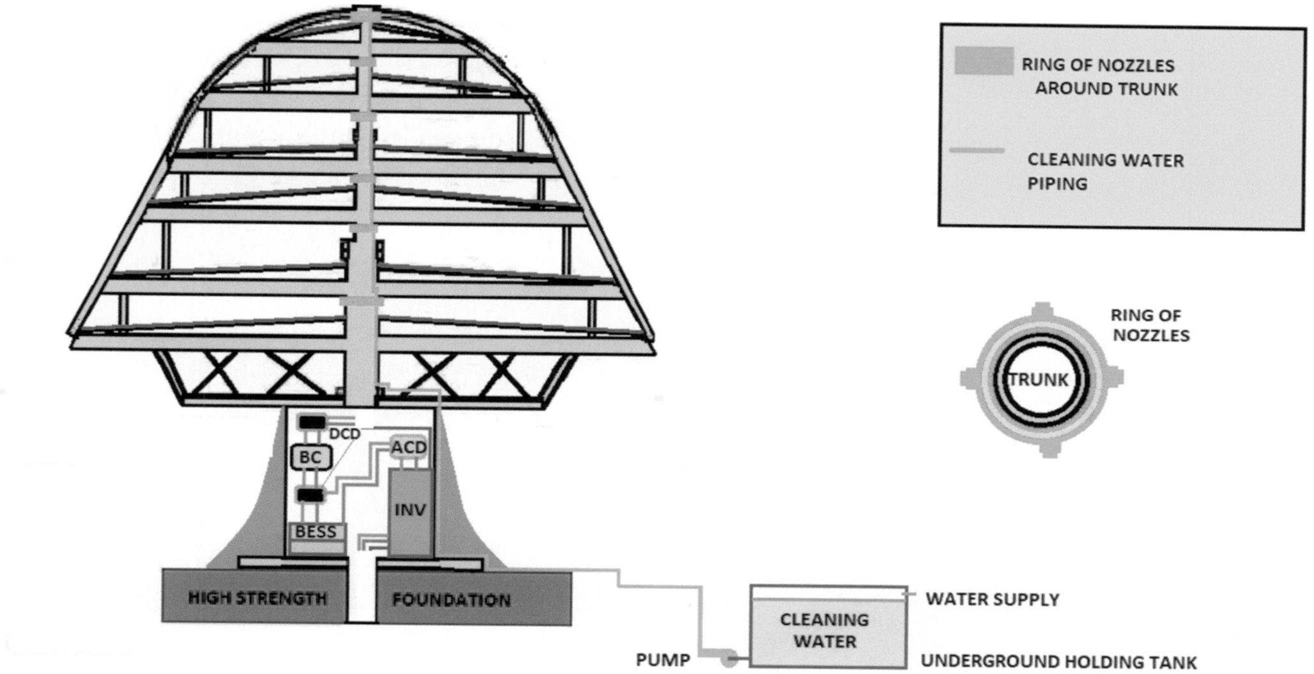

Fig 10.13 Typical Collector Cleaning System for NATARCH Solar Tree

The atmosphere contains significant amounts of dust and particulate matter at all times and this combination of dust and particulate matter usually settle on natural trees and on the surface of all existing PV systems collectors, thereby necessitating the regular removal of the particulate matter, which has the capacity to negatively impact the performance of the solar collectors. The collectors of the NATARCH solar trees will also be impacted by the dust and particulate matter in the atmosphere and this impact will be greater in dry arid regions where there are regular dust storms or when the NATARCH solar trees are utilized on a farm and acts as a barrier to soil erosion by wind. Most PV systems are currently cleaned using a fairly large amount of water and the NATARCH solar trees will require the same amount of water and a system to deliver it to the high surfaces of the spherical collectors. The components of this water cleaning systems shall include a water tank, pumps, nozzle spray heads to deliver the cleaning water to the surface of the sphere, a plumbing system that will deliver the water to the top of the sphere. Alternatively, scissors lifts may be utilized along with long hoses to deliver the cleaning water. The original design and Alternative 1 could utilize spray heads all around the circumference at all branch levels while alternative 2 design may utilize a deluge system installed at the top of the sphere.

Utilization of the Proposed NATARCH Solar Trees

The proposed NATARCH solar tree design could be utilized in residential, commercial, institutional, industrial and in a utility-scale operations, based on the desired load output and the space available for the installation. The utilization of the NATARCH solar tree could provide several benefits to each possible category as it leaves significant space underneath the canopies that could be utilize for many activities inclusive of agriculture, recreation, education, healthcare and any other possible use for available open land up to 15- 20 feet high.

Residential Installations

In single family residential units one NATARCH solar tree, depending on the surface area of the tree, could provide more electrical power than the home could utilize with enough left over to transfer to the utility grid, while providing other benefits inclusive of significant shade, eliminating the negative aesthetic and structural impacts of placing solar panels on each roof and also adding to the curb appeal of the property with the presence of a significant decorative piece. A typical residential installation is shown in Fig 10.14 below.

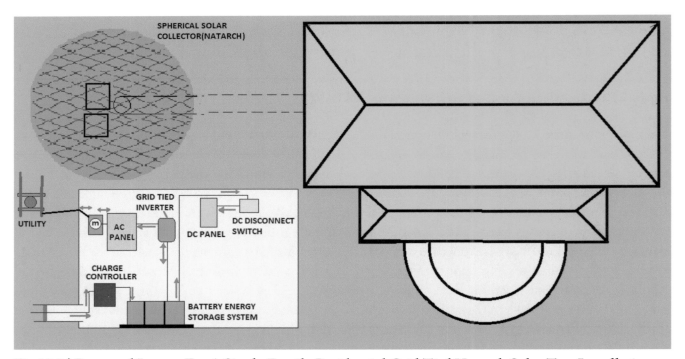

Fig 10.14 Proposed Layout For A Single-Family Residential Grid-Tied Natarch Solar Tree Installation

The NATARCH solar tree could also have significantly wider application on the residential front especially in regions of the world where wood is still a major source of energy for cooking and heating. The sourcing of wood for these activities has been established to be a major causative factor for forest fragmentation, degradation and deforestation in general, in many countries all around the world, as the poorest of the poor have no other source of energy that can be utilized for cooking and heating, major requirements for human health and life. The use of solar energy is spreading greatly in many poor areas of the world, however, the systems usually provided to these poor populations can only generate enough electrical energy to supply two bulbs at night, charging cell phones or maybe charging a laptop computer, but they never produce enough electrical energy for cooking and heating. However, the use of a NATARCH solar tree with a much larger surface area that can generate much more electrical energy could, with the help of transformers and other accessories, create enough energy to facilitate cooking and heating and make a major difference in the lives of people in these areas by supplying them with clean energy while at the same time reducing the pressure on the forests around them and also helping to conserve many species of flora and fauna that have been impacted by deforestation. The canopy of the NATARCH solar tree should also produce significant shade that could help with the reduction in the rate at which moisture is dried out of the soil surrounding most small villages and dwellings in the more rural and impoverished areas and the retention of more moisture should improve the capacity of the land to produce crops or feed animals.

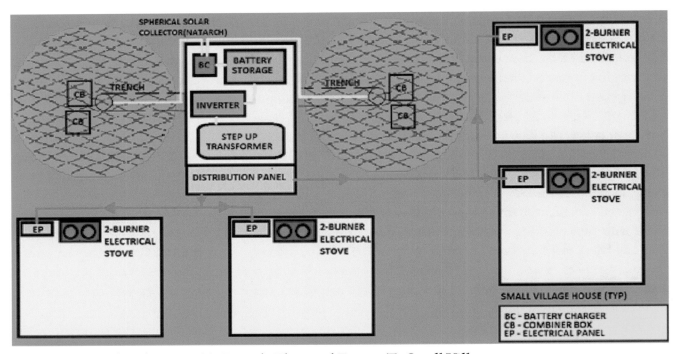

Fig 10.15 Natarch Solar Trees To Provide Electrical Energy To Small Villages

Combined Utility and Agriculture Installations

The utilization of NATARCH solar trees on the larger scale increases the benefits that could be achieved from one plot of land as the use of the NATARCH solar trees would leave significant spaces open that could be especially valuable in the circumstances where multiple utilization of the same land would allow the owner to increase their benefits from the land, as would be desired in the area of agriculture. Currently, most famers have to give up a large portion of their land, if not all, and their farming activities if they desire to benefit financially from utilizing their lands to earn an income from the solar energy business. A typical NATARCH solar tree field layout with agricultural activities could be as shown in Fig 10.16 below. The unused green areas could also be used for the planting of crops of many different kinds except for orchards where the trees are likely to grow tall enough to interfere with the light collecting equipment.

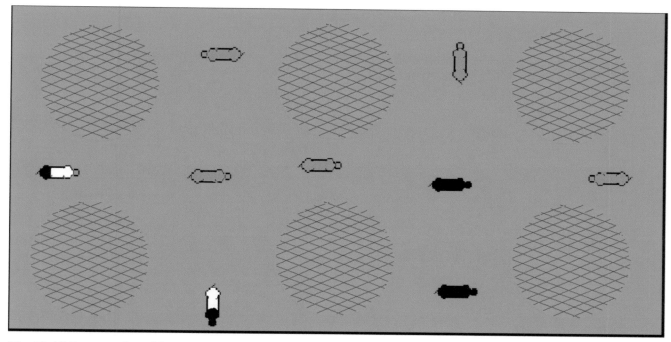

Fig 10.16 Proposed Field Layout Of Spherical Solar Collectors And Agriculture.

On a larger scale, an operation such as a dairy farm the benefits could be even greater and the combined operation could allow the farmer to increase his earnings significantly which would now be coming from two major streams of income. Apart from the direct financial benefit that the farmer could earn for getting involved in the solar energy business through the use of the NATARCH solar trees, the use of these trees could also provide other services to agricultures in the area of soil conservation as they could also be used, if properly laid out on a farm, to reduce soil erosion by wind, while simultaneously creating clean electrical energy. The canopy of the NATARCH solar tree should also produce significant shade that could help with the reduction in the rate at which moisture is dried out of the soil surrounding the farm and thereby also help to reduce the problem of soil erosion by the wind as the retention of more moisture should improve soil coverage with grass and increase the capacity of the land to produce crops or feed animals. Providing this soil conservation service, however, could significantly increase the amount of cleaning that would be required for the solar radiation collecting surface areas and increase the overall operating cost. In such a case, however, the water utilized would not be lost when it returns to the ground as it will provide life giving water to grass and other plants that are natural feed for animals.

NATARCH SOLAR TREES FOR TYPICAL UTILITY
SCALE POWER PRODUCTION

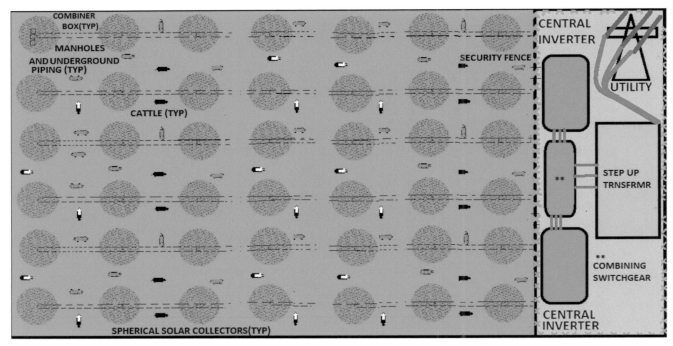

Fig 10.17 General Field Layout For Proposed Large Commercial or Utility Scale Natarch Solar Tree Farm And Cattle Farm.

The utilization of the NATARCH solar tree in other configurations could be in the form of multiples of the number of the trees that are highlighted in the diagrams above, starting with one for residential installations and growing to thousands for a utility scale application. This dual usage of the land could accelerate the rate of growth of the solar energy installations and this greater rate of growth would facilitate the required expansion in renewable energy supplies that is so needed to reduce the dependence on fossil fuels and consequently reduce the amount of carbon dioxide and other global warming gases that are being dumped into the atmosphere to meet the energy needs of our human population.

CHAPTER 11

The Future of Energy

The future of energy for human consumption must lay in the field of "nature energy" natural energy produced from the natural systems of the Earth, other than the fossil fuels which are also a part the natural systems, if humans are to escape the traps that the use of other fuel sources, inclusive of the fossil fuels, have always created, traps which become very hard to escape from when the economies of the world become intricately tied into the use of these fuels. Many of these traps are a natural consequence of the use of fuels that are in great supply and are available at costs that are considered to be economically affordable for most nations, irrespective of the long-term environmental costs associated with the use of these fuels. These traps that come with the use of these energy sources were developed very early in the creation of human societies as these early societies became more dependent upon these fuels and their relative proximity to the village and or towns, and these traps includes the depletion of natural resources, the creation of environmental problems, the creation of negative climatic conditions that impact the life of humans and the economic trap that is sprung when a community becomes totally dependent upon any one source of energy, especially a source of energy that is not renewable. The human need for energy is as basic as their need for food and water, a basic requirement for human survival and this is not a new requirement but has always been the case since societies started moving away from the earliest Hunter-Gatherer lifestyle that did not require a lot of cooking. The need for energy has grown greatly since humans started organizing societies around agriculture, a stable way of life which saw the human population living longer and producing more off-springs that ensured population growth. Later adjustments in the ways that communities were organized brought even more changes and these changes impacted how people worked, the creation of many more activities to occupy the time of humans and the development of these new activities required the growth in the amount energy needed to support the new activities.

In most early societies every family did everything for themselves and the usual complaint was that there were never enough hours in the days to complete the tasks at hand and this complaint would be most prominent during the reaping season when the families would have to do everything to ensure that the crops are reaped before the oncoming cold weather or other natural events that could be detrimental to crops. This meant everyone had to work late into the night using fuel to create artificial light by which to work and with this use of artificial lighting extending the number of workable hours, a practice that became a permanent part of each reaping season which would later extend to other activities. This demand for more energy has grown continuously with a growing world population, the growth in the number of human activities and with the creation of every piece of new technology that has occurred since the start of the industrial age, starting with the appliances in the homes and expanding to every piece of equipment in industry as industries had to grow continuously to meet the growing human demands for more goods and services.

The first energy resources utilized by humans included wood, grass, animals waste and solar energy and these energy sources were only used in very limited ways such as providing the necessary light and heat required in homes and for small scale processing of food products. This however changed greatly with the coming of organized villages and towns and later the industrial age which required vast amounts of energy to drive the new industrial machinery that were required to do the large-scale processing of food, the processing metal ores, the trimming of timber for the construction industry and later to create a new form of energy, electrical energy. The creation of this new energy, electrical energy also created a demand for more equipment to utilize it, a demand which saw the creation of many more technologies and machinery designed just to utilize this new energy, thereby creating an unending cycle for the greater utilization of more energy, a cycle which continue until today.

Unfortunately, the problems created by this never-ending cycle to utilize more energy has more often than not, never been a primary concern for those supplying the energy, nor those creating the new appliances nor the eventual end users of the energy and appliances. Historically, this group of suppliers and end users have never shown any concern about the source of the energy, nor the impact that the exploration, extraction, transporting, processing and the utilizing of these energies may have upon the environment of the Earth, that all humans must share, except in one case in Europe during the early days of the industrial revolution, when Europeans realized that they were running out of forests that were the source of the energy utilized to provide mechanical motion for their new factories. This realization forced the Europeans to develop the science of silviculture that sought to ensure that forests, their source of energy, were being regrown at a rate that was greater than the rate at which they were being used up by the factories, a practice that was not extended to the colonies later on.

This general attitude of a lack of concern displayed by most modern energy users and suppliers could be interpreted in two ways, a deliberate and general blindness on the part of most consumers or the callous manipulation of the world's populations by those who are preying on the most basic needs of the world's population for monetary gains only. This latter position, the manipulation for monetary gains only, is supported by many liberal economists and academics across the world as these economists and academics posit that the market driven capitalist economy is responsible for the continued use of fossil fuels and the worldwide drive to have more people utilizing more modern equipment that utilize electricity, thereby increasing the demand for more electrical energy and making their businesses more profitable. However, if one would stop to examine modern economies, market driven or planned, it would be seen that all of these economies have the same goal, growth, growth in production output, growth in the number of people employed, growth in the provision of services to the populations, growth in the production of profits or surpluses that will be beneficial to the shareholders or the general population who are said to be the shareholders in the planned economy. The path to achieving the desired growth in all the areas of an economy are the very same in all types of economies and this path basically involves the utilization of greater amounts of energy, energy that must be available and accessible at a relatively low cost, which means that Exxon Mobile is no more responsible for the continued use of fossil fuels and the emission of carbon dioxide into the atmosphere than are Gazprom or Sinopec, all very large oil companies which serve three very different types of economies. The high usage of energy to achieve growth would be the same even in the case in a purely planned communist or socialist economy that is seeking to grow, as all economic systems demand the same high usage of energy to achieve the desired growths in output to meet the needs of its population and to produce surplus for trade or profits for shareholders. Simply put, the units of energy to build the same size house, using the same material in China, Russia, India and the USA, is the same and the same applies to, equipment, machinery, and all things used in a modern economy, while the other inputs may differ.

The problem is therefore not the style of the economic system that governs in an economy, but the populations to be served and the creativity of the bright minds within each of those economies who will be creating more appliances, tools and toys to get the population to use more energy to meet their basic needs along with their social and recreational needs also. An examination of three of the leading economies in the world will highlight this high usage of energy as follows:

- US Market Driven Capitalist Economy. The US uses more energy per capita that the rest of the world, but not because it operates in a market driven capitalist economy, but because their economy created more automobiles, TVs, washing machines, water heaters, air conditioning systems and other high energy usage equipment per capita than the rest of the world at the current time. They also created more manufacturing plants and industries to create the goods and service needed in their country.

- China a mixture of a Communist Planned Economy and a Market Driven Capitalist Economy. China with a population that is more than four time larger than that of the US, has an economy that is only just as large as that of the US which is an indicator that there is great room for growth in the Chinese economy, an economy that has already grown at a very high rate over the last 40 years and is on a path to outstrip the US economy many times over in the next 50 years. To achieve their current level of growth China had to utilize and is still utilizing every low-cost source of energy available including the worst polluters of the environment even while they are also the leading users of clean energy technologies in the world. The high growth rate achieved by China while using coal and oil as the main sources of energy has made China the most polluted country in the world and this condition will only get worse as China seeks to elevate it very large population to a first world status, as they must continue using these sources of energy to achieve this goal. This first world status will see more Chinese driving cars, using more modern home and commercial appliances, living in modern homes with heat and air conditioning systems, along with a great growth in industrial activities to produce the vast volume of good and services required to ensure the desired growth.

- India with a mixture of a Market Driven Capitalist Economy and Planed Socialist Economy. India has the second largest population in the world but its economy is way behind those of the more developed world and China, but they also have the ambition to lift their large population to first world status which will see many more Indians owning and operating automobiles, utilizing domestic and commercial appliances, living in better homes that utilize both heating and cooling technologies and also growing their industrial activities to create more of the goods that their population will need. In order to achieve the rate of growth required India will also need to continue utilizing large sources of low-cost energy, inclusive of coal and oil that are among the worst environmental polluters in the world.

The USA, China and India all still use a mixture of fuels inclusive of fossil fuels, nuclear energy, solar, wind and hydropower, to drive their economies and each country has their own programs to reduce the amount of fossil fuels used before switching completely to cleaner renewable sources of energy, a switchover that may require another 50 years to achieve.

The same rules that apply to growth in the larger economies will also apply to the many other countries with large populations that are seeking to lift their people out of poverty, economic growth requires large sources of energy irrespective of the type of economic program that the country chooses to follow. This means that all of the countries of continental Africa, Continental South America, South East Asia and all large population centers of the world will soon be increasing their demands for large volumes of low-cost energy as they all seek to improve the living standards of the people of their nations in the next 50-100 years.

These real growth realities around the world clearly indicate that there will be a continuing demand for large volumes of energy and that these energy demands will be met by utilizing the lowest cost energy solutions available to them, solutions which will include the use of fossil fuels, nuclear energy and all of the existing renewable energy sources until the use of renewable energy sources outstrip that of fossil fuels. The other reality is that the growing demand for energy may soon create another major energy crisis as the world's ability to produce low-cost clean energy to meet growing energy demands while preventing the worst possible outcomes of climate change, may be stretched unless enough large non-carbon sources of natural energy had been studied, copied, synthesized and ready for utilization in the next 50 years. These sources of natural energy would include solar, wind, hydro, nuclear fission and fusion the regular renewables and lightening, fireflies, glow worms, the electric eel and every other natural technology which produces light, heat and or electrical energy, energy technologies which can be copied, synthesized and replicated for the clean production of electrical energy at the highest scale continuously.

To date great efforts have been made to ensure the development, deployment and distribution of the relevant technologies for nuclear fission, nuclear fusion, solar, wind and hydropower systems, all of which are fairly mature technologies at this stage, all of which are very important for the replacement of fossil fuels, but not enough is being done with solar energy which by itself still has the potential to meet the total energy need of the planet. Even with this vast, free and safe source of energy that is underutilized, much more effort has been dedicated to the areas of nuclear fission and nuclear fusion and very little attention is being given to energy produced by animals and plants, energy that could help to ensure that the populations could meet their potential of helping to eliminate the use of fossil fuels and creating a cleaner safer planet. Unfortunately, the study of nature energy seems to be limited to the areas of the major renewables and the nuclear energies, with the greatest interest been placed on the nuclear energies by nuclear physicists and those who value nuclear energy, energy that may be obtained by the manipulation of the atom. The rationale for the great support of the nuclear energies is of course that they produce large world-shaking forces and energies which can be harnessed for the benefit of all economies and while this is a fact, this ability to produce large sources of energy also produce large fears in the human population which hinders the widespread use of nuclear energy in places like the USA. An example of the failure of nuclear energy to achieve greater penetration is that over the years the results predicted for nuclear fission, the first of the nuclear technologies that was developed, have always fallen short of what was projected by its supporters, as these world-shaking sources of energy that were highly touted as an energy solution for the world came at a very high price, a price that most humans would rather not pay. At the current time only one country in the world that has utilized nuclear energy as their main source of energy (75%) France, while the developers of the technology only utilize nuclear energy to produce only 20% of the US energy demands (USDOE).

All current work in the area of nuclear fission is also quite controversial at this time as the use of nuclear energy is quite often used as the pretext for developing nuclear bombs for the destruction of peoples and nations, a position that should not be supported again by the international community and this position becomes a great barrier to the use of nuclear fission as a source of energy. Nuclear fission, the splitting of atoms, one of the greatest feats of science in the twentieth century, if properly harnessed could produce all the electrical energy that the population of the world needs, however, in its current state of imperfection, nuclear fission produces an extremely large volumes of nuclear waste that can be detrimental to the health, wellbeing and longevity of the planet and it is now deemed as a pariah among nations, a tools of world destruction in one form or the other. The truth is that while the splitting of the atom, nuclear fission, has produced a lot of energy, it has also split the world community as this same extremely large source of energy has also been used to produce bombs that can and have destroyed the lives of a large number of persons. This destructive ability was clearly demonstrated on two occasions soon after the technology was first developed, however as unfortunate as it may seem, this is not a new behaviour pattern among nations as on previous occasions when new energy sources have been developed, they have also been used for destructive purposes as well.

The development of nuclear fusion technology, the new holy grail of science, is now at the forefront of energy technology and with recent history in mind it is not impossible that efforts may be made to weaponize it, nuclear fusion, at some stage after the development process has been completed. The current hype around nuclear fusion clearly indicates that it will be promulgated as the new means to save the world, however, this new great solution may once again have the capacity to destroy the world, as fusion is the process that fires the stars, the greatest source of energy in the universe that humans are currently only getting their scientific teeth into and may yet be a long way before fulfilling the promise to provide clean safe energy to the world. It is therefore possible that the energy fate of the world may once again be placed on the shoulders of a source of energy newly synthesized by our nuclear physicists that is quite likely to bring with it the possibility that it may once again threaten to destroy us. Nuclear fusion with all of its great promises of providing a great source of energy, also comes with a great possibility of destruction and must never be accepted as the solution to all of our energy problems for life on Earth, until all bugs have been ironed out inclusive of any waste or other unsafe impacts, as it was meant for existence at least 93 million miles away from human, animal and plant life. As an alternative to nuclear fusion we must seek and find sources of energy that are compatible with life on earth only, green, renewable sources that are already a part of the natural environmental and ecological systems of Earth, energy from the plants, animals and the forces that are already a part of the creative forces within the envelope of Earth and the atmosphere, energy and forces that fostered and is a part of natural life on Earth, nature energy.

The development of energies for the future must therefore be focused on the current renewables energy systems and the lesser-known renewables and of course, any decision to develop these other nature energies would require another shift in the energy paradigm where the production of energy from the manipulation of atoms to produce extremely large volumes of energy with associated dangers, will give way to the production of energy from smaller, much safer sources. This shift in paradigms would require a move away from the atom of materials such as Uranium, Hydrogen and Helium and a simultaneous move towards the atoms of plants and animals that call planet Earth home and the atoms which cause lightening, thunderstorms and hurricanes. This change in focus would require that a new set of professionals be utilized as guides to finding new energy sources for continuing life on planet earth and these new professionals must be the people most closely involved with the plants and animals that call Earth home, botanists and biologists, and those closely concerned with activities within our atmosphere, atmospheric scientists.

Nature energy as defined by the leading thinkers in this developing field speak to all current renewable energy systems inclusive of solar, wind, hydro and bioenergy, but this category of bioenergy does not normally include natural bioenergy as can be seen in fire flies, only bioenergy as may be obtained from processing material such gaseous waste from waste treatment plants, wood, grass, corn, sugar, agricultural waste, and other plant waste to produce a fuel that may combusted to produce the desired heat energy and consequently to produce electricity. But in order to obtain the optimum results the scope of nature energy must be expanded to include the energy as produced by the fireflies and other animals, because the natural technology utilized by these animals, may be all very capable of helping the world to reduce and eliminate the dependence upon fossil fuels. A casual observance may suggest that the most obvious opportunity in this group is the light of the fireflies, cold light, light which is produced by bioluminescence within the body of the fireflies, a technology that has been proven over millennia to light up many dark nights without the utilization of any external source of energy or source of heat. This ability to produce light without heat or electricity is a natural technology that is partly understood but not yet mastered or replicated by humans and this natural technology, if properly harnessed could be utilized in eliminating the need to use electrical energy to produce necessary lighting in all sphere of life all across the globe. The elimination of the need to utilize electrical energy or heat to produce light could significantly reduce the total amount of energy required around the world as providing lighting utilizes as much as 20 percent of the worlds energy resources worldwide and this technology could also reduce carbon emissions by as much as 6 percent (USEIA). Second in prominence to the natural lighting produced by fireflies is the voltage produced by bioelectricity in electric eels (electrophorus electricus), electrical energy that these eels utilize to catch prey and which has been known to shock many human beings, as these eels produce voltages that are as high as 860 Volts and 1 ampere current (FAPESP 2019), all from their natural biological systems without any external inputs. This natural technology has been understood by humans and has thus far produced one technology that is currently very critical to renewable energy systems, battery technology, as it was said that Voltaire's first production of battery technology was based upon the electric eel (FAPESP 2019), however more work is still required on battery technology to reduce the possible environmental and longevity problems associated with them. As indicated by the examples of the fireflies and the electric eels above, animals produce lighting and electrical energy that if understood, synthesized and replicated could have the possibility to make a greater difference in energy usage by humans and thereby reduce the human need to utilize fossil fuels that harm the environment and consequently harm the lives of humans, animals and plants on planet Earth.

Unfortunately, the ability to bring these nature energy technologies into reality requires a large volume of resources inclusive of human resources, financial resources and time, resources which are normally dedicated to the more glamorous and attractive field of atomic science and the other areas of renewable energies inclusive of solar, wind and hydropower, areas which have always promised a greater return on each dollar spent in research and investment, two areas that have the potential to deliver vast amounts of clean energy. Unfortunately, while nuclear energy can and has delivered large amount of clean energy it is also associated with environmental and other dangers, dangers which have placed the global general public on edge for many decades and this fear of nuclear energy has limited the spread of its use across the US and many other areas of the world. Historical data indicates that nuclear energy research usually receives the largest share of the big dollars dedicated by the US government and the private sectors for energy research in the US. According to the US Department of Energy (DOE) for the period 1948 – 2018 the energy research budget of the department was as follows, renewables -13%, electric systems -5%, energy efficiency -11%, fossil fuels 24% and nuclear energy - 48%, the portions allocated to each sector have however, varied with time for the period from 1978 – 2018, showing small changes which are as follows, Renewables -18%, electric systems -6%, energy efficiency-16%, fossil fuels- 24% and nuclear energy- 37%. The changes shown here reflects a change in emphasis with renewables, electric systems and efficiency getting more attention but with nuclear energy still retaining the lion share of the research budget, indicating that the government still thinks that nuclear energy, with all of its inherent dangers and the fear that it induces in the public, still has the greatest potential to supply the greatest amount of energy, a position that is shared by many.

The other reality is that the research efforts in the atomic sciences does not always produce the results that many would have been expected based on the amounts invested and the public exposure given to these research efforts, especially the ones that were always in the eyes of the public, these high-profile projects includes things like the Large Hadron Colliders (LHC) in Europe and similar projects to study particle physics and subatomic particles, projects which do not appear to have produced the type of results that would have justified the money spent on them to the detriment of the development of things like natural energy as produced in animals, studies which may have produced much more productive results than such projects as the LHC.

The future therefore demands that more money be allocated to the development of the natural technologies that produces cold light, electric energy in electric eels and other fishes, electricity in lightening and also a shift in solar technology towards utilizing the natural architecture of trees. This reallocation of funding to support these natural technologies will of course impact the other areas of research, research that is still very vital going forward, however, based on the history of failures with some programs and the resources already allocate to nuclear science for such a long period of time, it could be beneficial to shift away small portions, 5- 10%, of current nuclear energy research budgeted funds to provide research funding for the natural technologies that could be just as rewarding and productive, if not more productive than nuclear science research which may have already reached it useful apex.

The field of organic solar cells for example is a clear indicator that we could probably be looking in the near future to growing our solar collectors instead of fabricating them from materials that are not conducive to good environmental health and wellbeing. The development of such a technology could also point the way to hybrid trees, NATARCH solar trees that will produce solar energy only, while providing some natural environmental services and allowing multiple uses of available land resources. More solar research alone, however, cannot bring about the required uptake to ensure that the desired growth in the usage of solar energy from the current level of 5% to the desired 40 % will be achieved in the required time window using existing or future solar energy technologies, unless the requisite policies and regulations are put in place by local, state and federal or national governments all around the world. These policies and regulation would include things like the mandatory use of BIPV systems on all new construction projects with a set time for all older building to be upgraded, in all sectors inclusive of residential, commercial, institutional and industrial facilities, starting at the roof and to include sidings, skylights, windows and all external components of a building that sees sunlight. Such policies and regulations could also bring a lot more research money especially from the private sectors, who would be designing, constructing and installing these systems on all building across the US and around the world.

REFERENCES

Solar System Retrieved from: https://cdn.teachercreated.com/covers/7633.png

Energy from the Sun-ACS Climate Science Tool Kit: Energy Balance Retrieved from:
https://www.acs.org/content/acs/en/climatescience/energybalance/energyfromsun.html

Sun and Earth Relationship Diagram. Retrieved from:
https://classconnection.s3.amazonaws.com/181/flashcards/1021181/jpg/1-142875952CF225F453C.jpg

Williams, M. (2014) Earth's orbit around the sun. Universe Today. Retrieved from:
https://phys.org/news/2014-11-earth-orbit-sun.html

EarthSky (2016) How long to orbit the Milky Way's center? Retrieved from:
https://earthsky.org/astronomy-essentials/milky-way-rotation

Our Solar System. Solar System Exploration, NASA Science. Retrieved from:
https://solarsystem.nasa.gov/solar-system/our-solar-system/in-depth/

Hanania, J., Sheardown, A., Stenhouse, K. and Donev, J. (2020) Nuclear Fusion in the Sun. Retrieved from:
https://energyeducation.ca/encyclopedia/Nuclear_fusion_in_the_Sun

Hocken, V. (2019) What is Earth's axial tilt or obliquity. Retrieved from:
www.timeanddate.com/astronomy/axial-tilt-obliquity.html

Liou (1980) Solar Constant. Retrieved from:
http://www.public.asu.edu/~hhuang38/mae578_lecture_03.pdf

Amin, S., Hanania, J., Stenhouse, K, Yyelland, B. and Donev, J. (2020) Solar energy to the Earth. Energy Education. Retrieved from: https://energyeducation.ca/encyclopedia/Solar_energy_to_the_Earth

Richardson, L. (2018) The history of solar energy. Energysage. Retrieved from: https://news.energysage.com/the-history-and-invention-of-solar-panel-technology/

Weiss, W. () Solar collectors. AEE INTEC. Retrieved from: http://www.crses.sun.ac.za/files/services/events/workshops/03_Solar_Collectors.pdf

Apricus (2016) ETC Solar Collectors Retrieved from: http://www.apricus.com/upload/userfiles/downloads/ETC_Collector_Overview_Int.pdf

Apricus Solar Products. Retrieved from: https://www.afcplumbingandsolar.com/wp-content/uploads/2018/04/Apricus-Evacuated-Tubes.pdf

GENI (2015) Solar electric and solar thermal energy: a summary of current technologies https://www.solarthermalworld.org/sites/gstec/files/story/2015-06-26/solar-energy-and-technologies-2015-03-11.pdf

Solar Hot Water Heater Thermosyphon. Pacific Northwest National Laboratory. Retrieved from: https://basc.pnnl.gov/resource-guides/solar-hot-water-heater-thermosiphon

Blackbody Radiation. Retrieved from: https://maths.ucd.ie/met/msc/fezzik/Phys-Met/Ch04-2-Slides.pdf

Bakari, R., Minja, R.J.A. and Njau, K.N. (2014) Effect of glass thickness on performance of flat plate solar collectors for fruit drying. Hindawi Publishing Corporation, Journal of Energy, Volume 2014, Article ID 247287, 8 pages. Retrieved from: http://downloads.hindawi.com/journals/jen/2014/247287.pdf

Solar Parabolic Trough. EERE Energy. Retrieved from: https://www1.eere.energy.gov/ba/pba/pdfs/solar_trough.pdf

IT Power India (2015) Material and components specifications: single axis tracked parabolic trough. IT Power India, Ministry of New and Renewable Energy, Government of India. Retrieved from: http://www.itpower.co.in/wp-content/uploads/2016/01/Final-Booklet-4-Parabolic-Trough.pdf

Dish/Engine System Concentrating Solar Thermal Power Basics. Office of Energy Efficiency & Renewable Energy. Energy.gov. Retrieved from:
https://www.energy.gov/eere/solar/dishengine-system-concentrating-solar-thermal-power-basics

Chapter 1, Understanding Sterling Engines in ten minutes or less. Retrieved from:
https://www.stirlingengine.com/download/9-12.pdf

Nakahara, H. (2008) Stirling engine. UBC Physics 420. Retrieved from:
https://people.ok.ubc.ca/jbobowsk/Stirling/pdfs/StirlingEnginePresentation.pdf

Farsakoglu, O. F. & Alahmad, A. (2018) Comprehensive design of Stirling Engine based solar dish power plant with solar tracking system. Department of Electrical and Electronic Engineering, Kilis 7, Aralik University, Turkey. Journal of Electrical and Electronic Systems. DOI:10.4172/2332-0796.1000248. Retrieved from:
https://www.hilarispublisher.com/open-access/comprehensive-design-of-stirling-engine-based-solar-dish-power-plantwith-solar-tracking-system

Butti, J. M. L., Rivier, S., Delgado, L., Rivera, S.S. and McLeod, J. E. N. (2018) Sun-tracking system design for parabolic dish solar concentrator. Proceeding of the World Congress of Engineering 2018, Vol 11, WCE 2018, July4-6, 2018 London, UK. Retrieved from:
http://www.iaeng.org/publication/WCE2018/WCE2018_pp447-452.pdf

IT Power India (2015) Material and component specifications linear Fresnel reflectors. IT Power India, UNDP, GEF, MNRE. Retrieved from:
http://www.itpower.co.in/wp-content/uploads/2016/01/Final-Booklet-6-Linear-Fresnel-Reflector.pdf

Jorgensen, G., Williams, T. and Wendelin, T. (1994) Advanced reflector materials for solar concentrators. NREL. Retrieved from: https://www.nrel.gov/docs/legosti/old/7018.pdf

Mustafa, M.A., Abdelhady, S. and Elweteedy, A.A. (2012) Analytical study of an innovated solar power tower (PS10) Aswan. Mechanical Power Engineering Department, M.T.C., Cairo, Egypt. International Journal of Energy Engineering 2012 2(6):273-278. DOI:10.5923/j.ijee20120206.01
http://article.sapub.org/pdf/10.5923.j.ijee.20120206.01.pdf

https://folk.ntnu.no/skoge/prost/proceedings/ifac2014/media/files/1581.pdf

NREL (1982) Basic Photovoltaics Principle and Methods Retrieved from: https://www.nrel.gov/docs/legosti/old/1448.pdf -(Basic Photovoltaic Principles and Methods (nrel.gov))

State Energy Conservation Office - Introduction to Photovoltaic Systems. Renewable Energy, The infinite Power of Texas. Retrieved from: https://www.austincc.edu/green/assets/seco-intro-to-pv.pdf

Photovoltaic (PV) Tutorial- Retrieved from web.mit.edu/taalebi/www/scitech/pvttorial.pdf

Dunlop, J. P. (1997) Batteries and stand-alone photovoltaic systems- fundamentals and application. Florida Solar Energy Center, University of Central Florida. Retrieved from:
https://edge.cit.edu/content/P14421/public/WorkingDocuments/power/Battery/battery_PV_system.pdf

Lane, C. (2021) What are the different type of solar batteries? Solar Review. Retrieved from: https://www.solarreview.com/blog/types-of-solar-batteries

IRENA(2019) Utility-scale batteries - innovation landscape brief. IRENA. Retrieved from: https://www.irena.org/-/media/Files/IRENA/Agency/Publication/2019/Sep/IRENA_Utility-scale-batteries_2019.pdf

WSU (2009) Solar Electric Systems Design, Operations and Installation. Washington State University, Extension Energy Program.
Retrieved from: www.energy.wsu.edu/Documents/SolarPVforbuiders2009.pdf

Sharp Electronic Corporation. Solar Power System Installation Manual.
Retrieved from: https://www.solarelectricsupply.com/media/custom/upload/sharp-all-installation.pdf

Pern, J. (2008) Module encapsulation, material, processing and testing. NREL. Retrieved from:
https://www.nrel.gov/docs/tu09osti/44666.pdf

Smalley, J. (2015) What is a combiner box? Solar Power World. Retrieved from:
https://www.solarpowerworldonline.com/2015/06/what-is-a-combiner-box/

Series and Parallel Wiring. Retrieved from: https://www.pveducation.com/solarconcepts/series-and-parallel-wiring/

MidNite Solar PV Combiners explained (MNPV).
Retrieved from: www.midnitesolar.com/pdfs/MidNitePVCombinersexplaineddiagrams.pdf

Jameel, D. A. (2015) Thin film deposition processes. Journal of Modern Physics and Applications Vol 1, No. 4,2015, pp 193-199. https://www.aiscience.org/jounal/IJMPA. American Institute of Science (AIS) Retrieved from: https://www.researchgate.net/publications/281090098 Thin Film Deposition Processes/ Link/55d44f3a08ae0a3417229683/download

Minneart, B. (2008) Thin film solar cells: an overview. University Gent.
Retrieved from: https://biblio.ugent.be/publication/4238935/file/4238983.pdf

Vonderhaar, G. (2017) Efficiency of solar cell design and material. Missouri S&T Peer to Peer 1. Retrieved from: https://scholarmine.mst.edu/peer2peer/Vol1/1552/7

Maxim Integrated Products (2015) Solar cell optimization: cutting costs and driving performance. 2015 Maxim Integrated Products. Retrieved from: https://www.maiximintegrated.com/content/dam/files/design/technical-documents/white-paper/solar-cells-optimization.pdf

Jim Dunlop (2012) Inverter – chapter 8. 2012 Jim Dunlop Solar.
Retrieved from: ecgllp.com/files/5614/02000/1304/8-Inverters.pdf

Franklyn, E. (2016) Solar Photovoltaic Systems Components. The University of Arizona, Colege of Agriculture and Life Sciences, Cooperative Extension (921742) May 2016. Retrieved from: https://extension.arizona.edu/sites/extension.arizona.edu/files/pubs/a21742-2018.pdf

https://www.alternative-energy-tutorial.com/images/stories/solar/altis.gif

Waterheatertimer.org/images/photo-voltaic-array-600jpg
https://www.elprocus.com/wp-content/uploads/2015/08/solar-charge-controller-using-microcontroller

Hybrid Source MISO Platform. retrieved from: https://www.cdnmisoenergy.org/

AE868 Commercial Solar Electric Systems: Inverter types and classifications. Retrieved from: https://www.e-edcation.psu.edu/ae868/node/904

Eme 812 4 Utility Solar Power and Concentration. retrieved from: https://www.e-education.psu.edu/eme812/node/737

https://vishalnagarcool.blogspot.com/2019/01/single-axis-solar-tracking-system-using.html

https://www.homemadecircuits.com/mppt-solar-tracker-difference/

https://www.researchgate.net/figure/A-diagram-illustrating-a-south-facing-solar-panel-PV-module-with-optimum-tilt-angle-figs3-333045245

Anatomy of a Rooftop Solar Mounting System. Solar Power World. Retrieved from: https://www.solarpowerworld.com/2014/03/anatomy-rooftop-solar-mounting-system/

Marsh, J. (2021) Solar trackers: everything you need to know. Energy Sage. Retrieved from: https://news.energysage.com/solar-tracker-everything-you-need-to-know/

Dutta, A. () Dual-Axis Drive System. Arindam Dutta. Retrieved from: https://www.researchgate.net/figure/the-overall-solar-tracking-system-based-on-resistance-sensitive-comparison-method-fig3-257910401

https://www.instructable.com/Arduino-solar-tracker-single-or-dual-axis/
https://solaryaan.com/modules-mounting-structure-types-basic-details/

MPPT. Retrieved from: https://www.solar-electric.com/learning-center/mppt-solar-charge-controller.html

PVeducator.blogspot.com/2015/02/how-offline-mppt-works.html

Patel, U., Sahu, D. and Tirkey, D. () Maximum Power Point Tracking Using Perturb & Observe Algorithm and Compare with Another Algorithm. International Journal of Digital Applications and Contemporary Research. Retrieved from: www.ijacr.com/uploads/papers/DhaneshwariSahu

Choudhary, S. and Jain, M. (Dr) A review of different maximum power point tracker (MPPT) approaches for PV systems. Oriental Institute of Science and Technology, Bhopal India. International Journal of Trend in Research and Development, Volume 7(2) ISSN: 2394-9333, www.ijtrd.com

https://www.researchgate.net/figures/4-Central-plant-inverter_fig1_309035631

mepits.com/tutorial/579/electrical/mppt-or-pwm-which-is-better
https://www,eecs.umich.edu/courses/eecs373/readings/dc-dc-primer.pdf

https://www.allaboutcircuits.com/technical-article/analysis-of-four-dc-dc-converters-in-equilibrium/

https://www.researchgate.net/publication/322780699.PDF

Tanomvorasin, P. and Thanarak, P. (2015) Availability of decentralized inverter concept of PV power systems in Ubon, Ratchathani, Thailand. School of Renewable Energy Technology (SERT) Naresuah University, Phinsauulok, 6500, Thailand. Retrieved from: www.sert.nu.ac.th/IIRE/login/FP_V10N2(5).pdf

Commercial Inverters (2019) Brochure. Retrieved from: https://www.wernernm.com/wp-content/uploads/2019/03/TOS-Commercial-Solar-inverter-Design-with-Solectria.pdf

Maleki et al (2017) Maximum and minimum value of declination angle. Retrieved from: Researchgate.net/figure/Maximum-and-minimum-value-of-declination-angle-Maleki-et-al-2017_fig3_324502848

Pauli, D., White, J.W., Andrade-Sanchez, P., Conley, M.M., Heun, J., Thorp, K.R., French, A. N., Hunsacker, D.J., Carmo-Silva, E. Wang, G. and Gore, M.A. (2017) Investigations of the Influence of Leaf Thickness on Canopy Reflectance and Physiological Traits in Upland and Pima Cotton Populations. Frontiers in Plant Science, Sec. Plant Breeding.
Retrieved from: https://www.frontiersin.org/articles/10.3389/fpls.2017.01405/full

Yates, M. J., Verboom, G. A., Rebelo, A. G. and Cramer, M.D. (2010) Ecophysiological significance of leaf size variation in Proteaceae from the Cape Floristic Region. Functional Ecology 2010. 24, 485-492. Retrieved from: https://besjournals.onlinelibrary.wiley.com/doi/epdf/10.1111/j.1365-2435.2009.01678.x

Zhang, Q., Hao, F., Li, J., Zhou, Y., Wei, Y. and Lin, H. (2018) Perovskite solar cells: must lead be replaced- and can it be done?, Science and Technology of Advance Materials, 19:1, 425-442, DOI: 10.1080/14686996.2018.1460176 Retrieved from: https://doi.org/10.1080/14686996.2018.1460176

Jannat, A., Rahman, M.F. and Khan, M. S. H. (2013) A review study of organic photovoltaic, International Journal of Scientific & Engineering Research, Volume 4, Issue 1, January-2013, ISSN 2229-5518. Retrieved from: https://www.ijser.org

Prezhdo, O. V. (2008) Multiple excitons and the electron-phonon bottleneck in semiconductors quantum dots: an ab initio perspective. Frontiers Article, University of Washington, Seattle. Retrieved from: https://dornsife.usc.edu/assets/sites/1066/docs/rev5.pdf

NREL (2013) Quantum dots promise to significantly boost solar cell efficiency. National Renewable Energy Laboratory, Denver, Colorado. Retrieved from: https://nrel.gov/docs/fy13osti/59015.pdf

Gu, X., Lai, X., Zhang, Y., Wang, T., Tan, W.L., McNeil, C. R., Liu, Q., Sonar, P., He, F., Li, W., Shan, C. and Kyaw, A.K. K. (2022) Organic solar cell with efficiency over 20% and V_{OC} exceeding 2.1 V enabled by tandem with all-inorganic perovskite and thermal annealing-free process. Advanced Science/ Volume 9, Issue 28/2200445 Retrieved from: https://doi.org/10.1002/advs.202200445

Paulo, S., Palomares, E. and Martinez-Ferrero, E (2016) Graphene and carbon quantum dot-based material in photovoltaic devices: from synthesis to application. MDPI, Journals/Nanomaterials/ volume 6/ issue 9/ 10.3390/ nano6090157 Retrieved from: https://mdpi.com/2079-4991/6/9/157/htm

Vercelli, B. (2021) The role of carbon quantum dots in organic photovoltaics: a short overview. Coatings 2021,11, 232. Retrieved from: https://doi.org/10.3390/coatings11020232

Kumar, P., Dua, S., Pani, B. and Bhatt, G. (2022) Application of fluorescent cqds for enhancing the performance of solar cells and wleds. IntechOpen. Retrieved from: https://intechopen.com/online-first/84613

Tang, Q., Zhu, W., He, B. and Yang, P. (2017) Rapid conversion from carbohydrates to large-scale carbon quantum dots for all weather solar cells. ACS Nano 2017, 11,2, 1540-1547. Retrieved from: https://pubs.acs.org/doi/pdf/10.1021/acsnano.6b06867

Hawkins, V., Szeltner, T. and Gallagher, M. (2016?) unique properties of the geodesic dome. Washkewicz College of Engineering, Cleveland State University. Retrieved from: https://www.csuohio.edu/sites/default/files/61A-TheUniquePropertiesofGeodesicDomes.pdf

FAPESP (2019) A new species of electric eel produces the highest voltage discharge of any known animals. FAPESP. Retrieved from : https://phys.org/news/2019-09-species-electric-eel-highest-voltage.html

Congressional Research Service (2018) Renewable energy funding history: a comparison with funding for nuclear energy, fossil fuel energy, energy efficiency and energy systems r&d. Congressional Research Service. Retrieved from: https://crsreports.congress.gov/product/pdf/RS/RS22858/17

Printed in the United States
by Baker & Taylor Publisher Services